AUGUSTINE AND EVOLUTION

AUGUSTINE AND EVOLUTION

A STUDY IN THE SAINT'S DE GENESI AD LITTERAM AND DE TRINITATE

BY

HENRY WOODS, S. J.

UNIVERSITY OF SANTA CLARA, CALIFORNIA

WIPF & STOCK · Eugene, Oregon

Wipf and Stock Publishers
199 W 8th Ave, Suite 3
Eugene, OR 97401

Augustine and Evolution
A Study in the Saint's De Genesi ad Litteram and De Trinitate
By Woods, Henry, S. J.
ISBN 13: 978-1-60608-688-9
Publication date 5/7/2009
Previously published by Universal Knowledge Foundation, 1924

PREFACE

Evolution, as understood today, is at the utmost little more than a hundred years old. How then can St. Augustine, who lived so many centuries ago, be called an Evolutionist? Certainly no one pretends that the Holy Doctor held or taught Evolution in any modern form. What is assumed is that his doctrine of creation finds its logical consequence in Evolution.

The question is purely domestic. It is raised by Catholics who wish to go a certain distance with the modern Evolutionist, yet perceive that a large number of theologians and teachers hold that the wish can be followed only at the sacrifice of a Catholic teaching and divine revelation. To justify themselves and to bring the more timid to their way of thinking they are eager to draw to their support the great doctors of the Church, and especially him who devoted himself to the interpretation of the scriptural account of creation in the literal sense.

The matter is abstruse enough to prevent many from investigating it. Nevertheless, the affirmation is made and repeated. It is then accepted under the impression that someone must have gone into the question thoroughly. With all due respect to those making the assertion, we have come to doubt the thoroughness of the work.

Hence, laying aside all private opinions and abstracting from the question of Evolution itself, we undertake the task, seeking under the guidance of the Angel of the schools to understand clearly the teaching of St. Augustine. The fruit of our toil will be found in our conclusions against the claim contained in the pages that follow.

We must add that, when in the citations no treatise is named, *De Genesi ad Litteram* is always to be understood.

University of Santa Clara, California.

CONTENTS

CHAPTER I

THE STATE OF THE QUESTION

CATHOLICS, holding modern ideas regarding the origin of existing nature, are anxious naturally to justify themselves before what they feel to be the accepted Christian teaching. By this they do not understand a creation lasting just one hundred and forty-four hours, beginning with chaos and ending in the groves and flowery glades of Eden, peopled with animals of familiar form, amongst which man walked supreme. While such an idea included no intrinsic impossibility, it would be unjust to suspect the Catholic Evolutionist of assuming that the whole question lies between it, and the theory he favors, that to reject the one, necessitates the acceptance of the other. The dispute is between fixed species and transitional, in the strict sense of the term, *species*, and between immediate and mediate creation as their origin.

The question in the last analysis is rather of fact than of theory. Not indeed that we can expect ever to bridge the vast expanse of time and, as it were, cross over to the nascent world, and see with our own eyes the origin of things in such a way as would settle the matter forever ; but in this sense, that the Holy Scripture contains facts of divine revelation, with which every

Christian theory must agree. On the other hand, the first chapters of Genesis are not a treatise on cosmic origins. Their object is not to give on these the plain statement of an astronomer, a geologist, a naturalist. They demand from this point of view interpretation, explanation. Hence, his natural anxiety for justification pushes the Catholic Evolutionist to seek support among those to whom we look for that interpretation, the Fathers of the Church. This he thinks he has found in St. Augustine's doctrine of creation in *"Rationes Seminales."*

The matter at issue, then, is, whether this doctrine contains the modern theory of evolution. That the discussion may be conclusive it is necessary to determine exactly what one affirming the question is absolutely bound to prove, in order to establish his position; and, consequently, what it suffices to show, in order to refute him.

Let us then begin by excluding what would cloud the issue. In the first place the Evolutionist need not prove St. Augustine to have been formally such as himself, nor his doctrine to have been formally evolutionary. Could this be established, the question would be ended. But not to establish it leaves the question still unsolved. Consequently, to show its contradictory, though it might be a useful preliminary step, would not reach the heart of the matter and settle the debate.

This is one extreme. Its contrary is to assume that if by giving the terms a sense drawn from Evolution and by conveniently limiting their application, one can twist the Saint's doctrine into some harmony with the modern theory, the Evolutionist may claim St. Augustine as his own. Here we must note that, if the opponents of Evolution sometimes fall into the former error, its supporters are not guiltless of the latter.

To claim an author's support for conclusions drawn from his doctrine, there is no need to show that he foresaw such conclusions, or that, had he lived to see them, he would have accepted them. For instance, knowing well that Modernism would have been utterly rejected by one whose orthodoxy is above all question, philosophers may hold, nevertheless, that in his philosophic system it finds support. To establish this they need only show that the system's principles, as understood and propounded by its author, lead logically to Modernism in their conclusions. But it is absolutely necessary to show that the conclusions flow from the principles in the author's sense, not in some sense attributed to them by his opponents, nor in that of those who, to claim his support, pervert them. So to make the doctrine of the *ratiures seminales* support Evolutionism, it must be made manifest that, as conceived and expounded by St. Augustine, they fit in perfectly with this theory. To show that it is Evolutionism undeveloped, requires the further demonstration that its development can lead nowhere else.

Wherefore, those who would draw from St. Augustine support for Evolution, must take his doctrine in his own sense simply and without gloss, neither grudging the toil without which it can not be penetrated, nor yielding to the temptation to read into it their own sense, and even to supply here and there the word or phrase necessary to the doing so. They must then show, at least, that its logical consequences are conformable to that system. It is our place to prove, after the same study, that St. Augustine's doctrine so understood has nothing that in any way favors Evolution.

CHAPTER II

THE OBJECT OF *DE GENESI AD LIT-TERAM* AND ITS DOCTRINE

THE doctrine of St. Augustine that gives rise to the present controversy, is worked out formally in his treatise, *De Genesi ad Litteram*. He had years before written a treatise on that book against the abuse of it by the Manicheans, in which he admitted that, so far at least as the ordinary interpreter was concerned, not everything could be interpreted literally. With this, however, he seems to have been dissatisfied. He returned to the matter again, and at last put forth the treatise mentioned in the title of this chapter. As the title implies, and as he himself tells us, its object is to interpret the book, not allegorically, but according to its proper historical sense.[1] Its aim, then, is not to account for the origin of things, for the process by which from matter without form, and therefore without external show, God perfected this world of varied beauty.[2] Neither is it to reconcile the sacred text with any philosophical system. He would not speak with the sure tone of a master who had worked out a theory of creation to satisfy, on the one hand, the searching investigation of the votary of natural science, and

[1] Retract., ii, xxiv, 1.
[2] Conf., xii, 4.

5

on the other, the jealous criticism of the guardian of supernatural revelation. He confesses that in the composition of the work more things were sought than found; and that, of what were found, fewer were established, the remainder being put down as still to be investigated.[8] In a word, he is speaking, not as a cosmologist to his fellows, for not to them did he submit the approving or the condemning of his work; but as a theologian subject to the correction of the Church.

For his attitude is quite clear. Given, say two possible interpretations of scripture, science is to be taken account of, only when it proves absolutely one or other to be untrue. Otherwise the determining of which is the literal interpretation is reached by investigating the intention of the sacred writer. This may be known from the context. If it be not, as is so often the case in these chapters of Genesis, the wise commentator gives all the approved opinions. It would be disgraceful to insist through ignorance on an interpretation that contradicts the certain truths of reason and experience. On the other hand, it would be outrageous for one puffed up with worldly learning to blame, as rude and unpolished, the sacred scripture, so manifold of interpretation for God's wise purpose, who gives it to serve man's manifold need. It would be base in a weak brother to yield to such arrogance, to fancy it a sign of prodigious learning, to come to loathe the holy text that should be his delight, and shrinking from the toil of the harvest field, to run after the flowers

[8] Retract., *loc. cit.*

of the thorn.[4] Evidently the holy doctor was thinking of anything but conciliating what we call science. His aim was no more than this, to contribute what he could to the determination of the literal sense of the history of creation, and in doing so to reconcile apparent contradictions in the sacred text.

The principal difficulty he offers to solve is the enumeration in the first chapter of Genesis of apparently successive creations on successive days, and the evident assertion in the second, of the accomplishment of the whole work in one day: "These are the generations of the heaven and the earth when they were created in the day that the Lord God made the heaven and the earth, and every plant of the earth before it sprung up in the earth, and every herb before it grew." [5]   Two problems confronted him—what was the nature of the days mentioned in the first chapter? and, how to explain the certain fact of the creation in a single day?

With regard to the first, he accepted as obvious, that days of twenty-four hours with their morning and evening, determined by the apparent motion of the sun were out of the question; since in the history of the six days the sun makes its appearance on the fourth only.   He therefore seeks the meaning of the term "day," in what makes the day, namely, light.   This decreasing is less clear at evening.   Growing it is brighter in the morning.   It is at its full splendor with the

[4] Lib., i, 38, 39, 40.
[5] Gen., ii, 4, 5.

noon. By it things are visible, obscurely in the evening, more clearly in the morning, at noonday in their full visibility. Visibility supposes its correlative, vision. To see creatures in the day of creation according to their various visibility, there were the angels. They saw them most clearly in the Word. They saw each in its own proper existence, obscurely in comparison with the bright vision in the Word. But this obscurity grew less and less, as the creature's relation to the Creator became more manifest, the evidence of a wisdom, goodness and power, compelling praise. These cognitions St. Augustine calls day-knowledge, evening-knowledge, morning-knowledge; their medium, the degrees of light, he holds to be the objects of the divine Author's terms, day, evening, morning.[6]

This explanation he proposes as probable, not as certain. It may not be the true sense. Others are free to think out another. He himself may find something to suit the text better. But there is question of *De Genesis ad Litteram*. Other explanations, whatever they be, must be propounded as literal in the same way as he offers his present exposition as literal. To call it allegorical would be, in St. Augustine's mind, a great mistake. Christ, he argues, is the light literally,[7] not figuratively, as when He is called the stone rejected by the builders.[8] Where light is better and more certain, there is the truer day; why, then, not a truer evening and a truer morning?[9] Note his

[6] Lib., iv, 43, 44, 45.
[7] John, viii, 12.
[8] Acts, iv, 4.
[9] Lib., iv, 45.

method of solving difficulties. He has recourse
to a spiritual light and a spiritual day; not as
passing from the literal to the allegorical, but as
from the less real because participated, to the
more real because absolute. With perfect right
he seeks in the Creator the literal interpretation
of the Creator's own word. But it is a method
that hardly affords hope of finding in the *rationes
seminales* anything akin to a theory so material as
Evolution. This, in the Catholic mind that accepts it, protects God's honor by excluding Him
as far as possible from His creature.

Coming to the seminal reasons in connection
with the creation of all things "in the day that
the Lord God made heaven and earth," we observe
that St. Augustine saw the absolute impossibility
of any successive creative acts on the part of God.
Moreover, if the six days were not natural days,
as we know them, there is certainly no question
of such a day in this one day. On God's part
creation could only be one single act identical with
the simple essential act that is Himself. "He
spake, and all things were made," [10] "creating them
simultaneously." [11] Yet nothing is more certain
than that the beginnings of created existence were
successive. These two facts St. Augustine might
have reconciled legitimately by distinguishing, as
does the scholastic theologian, God's intrinsic act
in His eternity, and its extrinsic term in time.
That he did not do so, can not be ascribed to
ignorance. That great mind could not have ig-

[10] Ps., cxlviii, 5.
[11] Ecclus., xviii, 1.

nored a distinction so elementary; the less so, because it is included virtually in his specific doctrine. But just because it is elementary, too near the surface of things for one who dared, as far as mortal might, to penetrate to the inner things of the Creator, he passed it by, to go deeper down to creation's very roots. His solution, then, if not of scholastic theology, is still less of modern science. It is rather allied to the unutterable things of the third heaven, which St. Paul, though he had seen them, could not express. "God created all things simultaneously in the term of His act, by imposing upon elementary matter the seminal reasons of all things that were to appear in the long course of time." What are these seminal reasons? For the Evolutionist the temptation to identify them with his theory is too great. It all seems so easy, so obvious. This should rather lead him to hesitate. St. Augustine was the last man to be satisfied with the obvious. He is dealing with solemn obscurities, the divine operation, the divine scripture telling of that operation, the divine sense of that scripture, the divine intelligence conceiving both the operation and its history, the divine will expressing both. Had he had any conception corresponding to any merely natural theory of origins, he would have explained it, the easy and the obvious, in a few words. To the seminal reasons he devotes chapters; and in the end bows down his intellect before their mystery, not as unequal to the investigation of the natural, but as overpowered by the immensity of the divine.[12] His mind was formed to rise spontaneously to the divine. Let us, then, add to our necessary preliminaries a brief study of that mind; since, unless it be sufficiently understood, it were vain to grasp at its doctrine. Almost unique in its own age, at least in degree, its kind is unknown today amongst men of science. Hence, the study of it is really necessary.

[12] Lib., vi, 10, 11.

CHAPTER III

THE MIND OF ST. AUGUSTINE

THE atmosphere of positive theology in which St. Augustine lived differed greatly from that of modern experimental science. In it he, who by the trend of nature was led to philosophical speculation rather than to physical investigation, was a theologian in the strictest sense, giving himself not so much to divinity as to the divine; and this constantly, not on occasion, making it the business of his life, the food of his soul, the matter of his continual meditation. He occupied himself with God rather than with creatures. Everywhere God met him. God's operation in the creature was to him much more than any mere physical theory of the creature's mode of origin. We must, therefore, be prepared to find in him thoughts too deep to be understood in a moment by those breathing another air; thoughts which even those who stand before him as disciples before their master must ponder long to reach their full meaning. The profundities of: "In the beginning God created the heaven and the earth," [13] of the Old Testament, with its complement from the New: "My Father worketh hitherto and I work"; [14] the mysteries of the six days of creation

[13] Gen., i, 1.
[14] John, v, 17.

taken in connection with the one day in which the Lord made heaven and earth, so filled and overwhelmed his intellect as to leave little opportunity for expatiating formally on genus and species. The whole tone of the treatise: *"De Genesi ad Litteram"* is of one walking in the obscurity of the supernatural, seeing in a glass darkly the secrets of the divine operation, not of one treading confidently in the light of nature the paths of natural science. *"Mira profunditas eloquiorum tuorum"* the cry had been wrung from him before by the meditation of these things. *"Mira profunditas, Deus meus, mira profunditas. Horror est intendere in eam; horror honoris et tremor amoris."* [15] And with the lapse of time the loving horror grew. He was swooning in the apprehension of the Creator.

And for this his life even before his conversion prepared him. Then he was but another example of what was not infrequent in the old pagan world, a body held fast in the bondage of concupiscence yet containing a soul capable of soaring towards the ethereal regions of pure intellectual contemplation. A sensualist and, at the same time, a philosopher, he came upon Plato as interpreted in the Alexandrian schools, while drawing near his emancipation; and welcomed the master's doctrine, approaching so near the truth, yet missing it so disastrously, that all this lower world contains is but an imperfect realization of the absolute realities, the substantial ideas of a higher state of being. Such philosophy, making any

[15] Conf., xii, 17.

thought of the things of our earthly existence
tame and incomplete, encouraged its disciples to
fix their attention on a future state, after the puri-
fication of the soul from the impediments of sense,
begun during life, shall have been perfected by
death, a state in which shall be seen the very truth
of the mere shadows that occupy mankind in this
present existence. It predisposed him to meditate
sublimely after his conversion on reality, absolute
and universal in the divine Essence alone, relative
only and participated in creatures. Thus he came
to see wonderfully how, as the outward expression
corresponds to that inward concept which contains
all the tongue can utter, so all creation is contained
in the Word, the interior substantial concept of the
divine Intelligence, the Second Person of the
adorable Trinity, by Whom, according to special
attribution, all things were made. Hence he perceived that these imitating in their actual existence
this divine Word, imitate the divine Essence of
which It is the express image.[16]   This is the
foundation of all his study of creation. Throughout his treatise he does not lose sight of it.
Having fixed his mind on God, rather than on
the creature, he views the creative act primarily
in God, secondarily only in its term. The act in
God is absolutely simple, therefore in that act
were created *"omnia simul."*

Thus, *omnia simul* is, as it were, his keynote.
He conceived creation as proceeding from its
Creator, a unit including all things whatsoever
that are to exist to the end of time, and corre-

[16] Lib., i, 9; Lib., v, 29.

sponding to the single creative mandate. Having
as his foundation all creatures as yet non-existent
in themselves, yet existing in God in their exemplary ideas as the object of divine knowledge,
he places the analogue of all things, as yet without individual existence, existing in elementary
matter as forms in potency, forms decreed to
exist, therefore no figments of the mind, no mere
entia rationis, but, as he terms them, *rationes
seminales*. They are the reasons, distinguishing
objectively the things that are to be, from mere
possibilities never to be actuated. They are seminal, not seeds yet following the analogy of seeds;
because, impressed on matter, they determine its
potency to what is to exist, and exclude all other
possibilities, as the virtue of the seed determines
matter to this species and excludes that. They
are realities so strictly, as to be the objects of
the angels' evening knowledge, the motive of their
morning praise. Nothing material that is to exist
can escape them. "As mothers are pregnant of
their children, so the world is pregnant with the
causes of things coming to birth." [17]   On the one
side, the Saint sees God, His divine Essence in-
finitely imitable, its actual imitation formulated in
exemplary ideas. On the other, the creature,
matter, in the abstract infinite in potency, its actual
potency formulated in seminal reasons. A paral-
lelism worthy of an intelligence penetrating
deeply into mysteries, but a long, long way from
natural history and physical science.

[17] De Trin., iii, 16.

CHAPTER IV

PRIME MATTER

TO follow St. Augustine in his explanation of the divine revelation, we must understand three things as he understood them, namely, prime, or unformed matter, time, eternity. We must note that, though for convenience sake we are going to employ scholastic terminology, we are not ignoring the fact that St. Augustine takes unformed matter in the scriptural sense for matter without definite external form. Yet in this we do no violence to his doctrine. St. Augustine agrees with every scholastic, that the external specific form must find its reason in the interior substantial determinant, we call the substantial form.

This calls for no proof. It is evident from his whole treatise. His doctrine then begins with the text: "Thy almighty hand which made the world of matter without form." [18] This is to say, of matter considered as absolutely undetermined, a pure passive potency. Nevertheless, this "matter without form" is neither coeternal with God, as if made by none, nor did another make it that God might have material from which to form the world.[19] But this does not touch our question.

[18] Wisdom, xi, 18.
[19] Cont. Adv. Leg. et Prophet., i, 11.

16

This begins with, What is prime matter in itself? What it is relatively, is clear enough, namely, the primary determined principle, purely passive, of every material substance, as the form is the primary determining principle, purely active. But one who had undertaken the task of explaining, as far as God gave him light, the literal sense of God's revelation, that of this His creature He had formed the world, could not rest content with the mere relativity of what God expressed in the term, "matter without form." What is its reality? It is not nothing, that is evident. It is not something in the way that every existing thing is something; that is equally clear, since for such "something" is needed the determining form. As Aristotle observes, it receives none of those categorical predicates which follow necessarily material substances. We cannot say of it, "What thing," nor "how great," nor "of what kind"; [20] since all these suppose being substantially complete. Therefore, it holds a middle place. But, says St. Augustine, these are mere words drawn by the mind from a spirit full of corporeal forms, and in his arduous task he was struggling to rise above the corporeal. "Could we say: 'Nothing something,' 'Is-not is,' I would call it this. And yet it already existed somehow, so as to receive these visible and composite shapes." [21] Elsewhere he illustrates as follows: A confused shouting is a noise without words. When formed into words it becomes articulate speech. Thus

[20] Metaph., vi, 3, 4.
[21] Conf., xii, 6.

the cry is formable, because it receives the form: the word is formed, because it has the form. Words then are formed from the voice which, though unformed, is not nothing.[22]

Such prime matter, nevertheless, can exist only under some form. "We must not think of God as first creating matter," the Saint admonishes, "and after an interval of time giving form to what He had created without form; but as creating it simultaneously with the world. As spoken words are produced by the speaker, not by giving form afterwards to a voice previously without form, but by uttering his voice fully formed, so we must understand that God did indeed create the world from unformed matter, yet concreated this matter simultaneously with the world. Still not uselessly do we tell, first that from which something is made, and afterwards what is made from it; because, though both can be made simultaneously, they can not be narrated simultaneously." [23]   This we find again in the treatise we are especially discussing. "When we say matter and form, we understand both simultaneously, though we cannot pronounce them simultaneously. As in the brief space of speaking we pronounce one before the other, so in the longer time of narration we discuss one before the other. Still God created both simultaneously, while we in our speech take up first in time what is first in origin only." [24]

Prime matter can be called not only what it

[22] Cont. Adv. Leg. et Prophet, *loc. cit.*
[23] *Ibid.,* 12.   [24] Lib., i, 29.

actually was under some elementary form, but also what it was to become by future formation. This most important principle St. Augustine lays down in explaining against the Manicheans the text: "In the beginning God created heaven and earth." He says: "Unformed matter is here called heaven and earth, not because it was this, but because it was able to become this; for heaven, it is written, was made afterwards. For if, considering a seed, we say that roots and wood and branches and fruit and leaves are there, not because they are there now, but because they are to be from it, in the same way it is said, 'In the beginning God made heaven and earth,' as if he made the seed of heaven and earth, when the matter of heaven and earth was still confused. But, because heaven and earth were certainly to be from it, matter itself is already called heaven and earth. Our Lord Himself uses this manner of speech when He says: 'I will not now call you servants, because the servant knows not what his master does. But I have called you friends, because all things whatsoever I have heard from the Father, I have made known to you.' [25] Not that he had actually done so as yet, but because the manifestation was certainly to take place." [26]

This predication of the future as past is not intended to denote a mere extrinsic attribution in the past, by the way, as it were, of prophecy. It expresses an intrinsic determination of the

[25] John, xv, 15.
[26] De Gen. cont. Manich., i, 11.

potency to what was to be. So St. Augustine says: "Unformed matter was called heaven and earth, not because it was so, but because it was able to become so." That is to say it had its potency determined to this. So our Lord would say: "I will not now, at this present moment, call you servants because there exists in you, declared when I called you to the Apostolate, a positive determination of your potency to receive my full revelation hereafter, and this is your present title to be called friends." This determination of the potency comes from divine providence decreeing the whole order of existing creation; and so St. Augustine says explicitly: "For what is called the unformed matter of things, capable of receiving forms and subject to the operation of the Creator, is convertible into all things, which it pleased the Creator to make; nor was it before those things which are seen to have been made from it.[27] In the beginning, therefore, God created prime matter with its potency positively determined to all things that were to be, so that these things may be said literally, not figuratively, to have been created simultaneously with it.

This distinction between abstract potentiality and concrete, perhaps not sufficiently adverted to, is of universal application. In the first moment of existence every concrete being is speculatively and in the abstract in potency to every act possible to its specific nature. But such universal potency would be universal impotence in a contingent being. A creature equally ready for every

[27] Serm., ccxiv, 2.

abstract possibility, contradictories and contraries included, would at any moment be really ready for nothing. For any concrete actuality an initial determination of potency is needed. This determination comes in a large degree from circumstances. This new-born babe is speculatively in potency to be Emperor of China, a bandit, a modern Shakespeare, Archbishop of Canterbury, an acrobat, and other things in infinite variety. His circumstances will restrict that abstract potency to the concrete potency of a farm, a place under government, a merchant's office, medicine, the law, etc. As he grows up, he will continually restrict his potency by his own free choice and the consequences of his own free acts; so that at the end of his life he shall have actuated it by a series of definite acts. Let us suppose now that his parents' function included the prevision of all those acts in all their freedom as far as they were free, and according to that prevision to provide by the act of generation suitable matter in which they were to be performed. They would have to proportion the receptivity of the matter to the act, or, in other words, to restrict the abstract potency to a concrete potency to these acts and no others, so far as this particular individual would be concerned. Now this was just the relation of God to his rational creatures in that providence in which, after seeing what would be the course of things in every possible creation, He decreed the actually existing order. What is true here of the individual, is true also of the race, and of irrational and inanimate being. Going back then to the first creation of matter, it was in concrete potency to all those acts, and to those only, which were to actuate it down to the end of the world. Its universal potency, a negative order to all possible material beings, became a positive relation to all actually future material beings by a positive intrinsic determination received from the divine will, that in creating it, created it for the actual order of creation which the Creator had chosen.

This positive determination of the abstract potentiality of prime matter in no way changes its purely passive nature. It remains such as St. Augustine conceived it when he said he would gladly call it "Nothing something," or "Is-not is," were such terms possible. This point is crucial. To suppose a change to active potency would suppose a contradiction, namely, prime matter actually existing under the future form, whatever this might be; for without form, the active principle, there can be no active potency. Nor can it be urged against us, that since definite prime matter becomes by this positive determination capable of all its future forms, and these are far more frequently the immediate effect of natural forces than of creation, the determination in question must introduce active potency. St. Augustine understood this clearly enough, as we shall see later. Nevertheless, he saw most clearly that all these natural forces, coming, as they do, from specific forms, originated with creation of those forms in prime matter. When, therefore, pushing the question further back and asking:

What is this prime matter in itself and how does it come to receive this definite form and not another, he perceived, as we also must perceive, that a complete answer required the determination of the passive receptivity of prime matter to the reception of this definite form. For as matter of its nature precedes form, its passive determination must precede the activity that comes from form.

Nor does the fact that after creation came the production by natural agents do anything else than confirm this conclusion. Were there question of creation only, not of the conciliation of simultaneous creation and the successive existence in time of its effects, such a positive determination of passive potency might, perhaps, be superfluous. But whatever be said of creation, once God decreed to add to it existence by natural production, the positive determination of passive potency to the things that were so to exist became imperative. Universal passive potency must be negative. It is the very opposite of the act by which the Creator moves his creatures, without which no action is possible. This is *positively* universal, moving the creature in its native tendency to its own good, with a positive generic motion including all acts now concretely possible, to this, without excluding that, and to that without excluding this. Universal passive potency shows nothing of the sort. It can be anything simply because it is absolutely indifferent to everything. But, unless there be an agent capable of determining it to this particular form it must be nothing. A created material agent has its activity. It can produce its like provided its activity be reproductive. This, however, supposes three things, the active form, actual participation by its own matter in the activity of the form, and responsiveness of external matter to that activity. This responsiveness of passive matters is essential. We may speak biologically of vital force; but this belongs to the living agent. It cannot reach out to change elementary matter into living substance, unless it meets in that matter a disposition to receive the living form. And this disposition can be no latent force, no communicated activity. These would but push the question back. The mystery of reproduction is the mystery of creation. The disposition of matter to life is simply and necessarily the determination of the abstract potency of this definite matter by the Creator's Word to become in its time and place this particular organism, not through some hidden physical force, but solely in obedience to Him, "Who spoke and *all* things were made."

This it was that, firmly grasped, commended to St. Augustine an argument which for him never grew stale, though we, perhaps, do not always catch its full strength. Why, he asks again and again, why wonder at our Lord's miracles? Why hesitate, when the same Lord works the same miracles around us every day? We are amazed at the water changed to wine during the wedding-feast; yet the same is done year after year in the vines, and we take it as a matter of course. The recalling to life of the dead who had lived fills us with admiration; the daily birth of those who before were not leaves us unaffected. These works of nature are held cheap by men, not because they are easy, but because they are constant.[28] Again on the miracle of feeding the five thousand, he remarks that, because we do not see God, and because the miracles by which He rules the world and conducts nature in its course have become commonplace, hardly anyone deigns to attend to the marvels contained in the meanest seed.[29] Elsewhere he repeats the argument from wine and seeds,[30] and that of resurrection and birth;[31] while such formulas as: "Not the cheapness, but the frequency," [32] and: "Not because it is greater, but because it is rare," [33] regarding respectively the wonder of nature and the formal miracle, are frequent in their appearance.

It may be said with perfect safety that to few does such an argument appeal as it did to him. The ordinary mind answers incontinently: "There is an essential difference between the works of nature and miracles. There the laws of nature work their work without interruption. Here we see exceptions to them and effects contrary to their operation." This is quite true; and St. Augustine would be the last to deny the definition of a miracle. His argument against the incredulous was: "Why deny the miracle because it

[28] In 2 John, Tract., ix, 1.
[29] In 6 John, Tract., xxiv, 1.
[30] Serm., cxxvi, 4.
[31] Serm., ccxlii, 1.
[32] Serm., cxxx, 1.
[33] In 6 John, Tract., xxiv, 1.

calls for the exercise of almighty power, when you see in nature so many unbroken series of operations calling for the same power?" The miracle depends upon a will and power supreme over nature. Nature is what it is, only by that supreme power and by virtue of the decree of that supreme will. Examine the seed. What is it? Matter under a certain form; the same matter in kind that is found in other seeds, in earth, water, air. It came into existence by generation from a plant of the same kind, and within that little thing is the power of producing another such plant. How is this effected? You say, by vital force coming from the form. But I must ask, how could the form of the parent plant by its vital force make the matter of this seed participate in that force, so that, as a substantial element under the activity of its form, it will take other matter from the elements, and make it capable of sharing in the same vital force? How does matter become obedient, in this way in one case, in that way in another, in a way utterly different in a third, and so through all the diverse natures of material being? What fixes the limits of this receptivity of matter? These are the questions that occupied St. Augustine's mind. These he would answer. The scholastic philosopher, consistently with his own function, replies that all material forms exist in the potency of matter, whence they are brought forth by the activity of other forms using physical forces and agents as instruments. St. Augustine would go further. He is not engaged in Cosmology, but in investigating the creative act, so as to reach the scripture's literal sense. What is this potency of matter? No full answer can be drawn from the mere processes of generation. We must seek it in the origin of things; and see the same infinite creative power, that dominates matter in nature's laws, determining the abstract passive potency of prime matter to its concrete potency with regard to all things that are to exist, making its actual receptivity adequate both to nature's laws and to the exceptions to them decreed by divine providence, that is, to natural processes and to miracles.

Hence he writes: "We refuse the name of creator, not only to the husbandman, since we read: 'Neither he that planteth, is anything, nor he that watereth, but God that giveth the increase'; [34] but even to the earth itself, fruitful mother, though it seems to be of all things, bringing from seed what springs up, and containing what is fixed in it by the roots, since we read in like manner: 'God giveth it a body as He wills, and to every seed its proper body.' [35] So, too, we ought not to call a woman the creator of her offspring, but Him rather who said to a certain one, His servant: 'Before I formed thee in the womb, I knew thee.' [36] And although the soul of one pregnant, if it be affected this way or that, can clothe the fetus with certain qualities, yet such a one does not make the nature produced, any more than she has made herself. In

[34] 1 Cor., iii, 7.　　　　[35] *Ibid.*, xv, 38.
[36] Jer., i, 5.

the generation of things, therefore, whatever corporal or seminal causes are employed by the operations of angels or of men, or of certain animals, or else in the mixture of male and female elements; also whatever effects in features and colors, the desires and movements of the mother's soul are able to produce in the tender and yielding unborn babe, no one is absolutely the maker of these natures affected thus or otherwise in their kind, but the supreme God whose hidden power, penetrating all things by His presence, causes to be, whatever exists in any way, inasmuch as it exists; because unless He made it, it would not be such a thing, or such another, but would be utterly incapable of existing." [87]

One content to remain on the surface of things might assert that St. Augustine had in mind only the obvious truth that nothing can come into existence without God's conservation of the agents, and His coöperation in their acts. For the rest, what he says about seminal causes may fit in very well with Evolution. This would be hardly complimentary to the great Doctor whose support Evolutionists are so eager to obtain, since it would be to make him one of the most striking examples verifying:

Parturiunt montes, nascetur ridiculus mus.

The more modest and more prudent course would be to hold with us, that St. Augustine's habit was to go below the surface, deeper than most men, into the realities of things.

[87] De Civ. Dei., xii, 25, post med.

Besides, St. Augustine is not discussing conservation or coöperation, but creation; and that in its primary origins. Moreover, the question, as we have said, is not, whether Evolutionists, after having put their own sense upon his words, can fit them in with their theories, but whether in their author's sense they are intended to express, even in the most rudimentary manner, such a theory or, at least, lead to it necessarily. Certainly, had the holy doctor had in his mind what Evolutionists suppose, he wasted much time and thought, not to say ink and paper, over what he could have said in a brief paragraph. For, though we are not discussing the evidences of Evolution, we may, perhaps, take the liberty to say that it has this in common with many other modern theories, that its expression has all that clearness that comes from lack of depth. How far St. Augustine was from such theorizing, and, at the same time, how far such theorizers are from understanding him, let the following show: "I was more ready to opine that what is without form ceases to be, than to conceive something between formed and nothing, neither formed nor nothing, unformed almost nothing. And my mind ceased to question my spirit, full on this account of images of formed bodies, changing them and varying them at will; and I fixed my attention on bodies themselves, and looked more deeply into their changeableness, whereby they cease to be what they have been, and begin to be what they were not. And I suspected that this passage from form to form was made by something unformed, not by absolutely nothing. But I wished to know, not to suspect. And should my voice and my pen confess to Thee all whatsoever Thou has made clear to me on this question, who of those reading could sustain the grasping?" [38]   In all sincerity, then, was St. Augustine striving to express the easy theorizing of evolutionary force, or the mystery of determined passive potency as lying at the foundation of creation?

[38] Conf., xii, 6.

CHAPTER V

TIME AND ETERNITY

WE now go on to investigate St. Augustine's concept of the relations between time and eternity. The matter is profound; and into its profundity that great mind penetrated deeply. Nevertheless, his word is as clear as his doctrine is deep. It calls for patient meditation. But it yields this reward, the conviction that it cannot lead to modern Evolution.

Time is the measure of corporeal motion. From this arose many questions in the Saint's mind, which, as they do not touch our study, we may omit. One conclusion, however, is of great importance, namely, that without a material being, or, at least, without a being in constant continuous motion, there could be no time. Time, then, is inseparable from the material creation.[39] Hence, not only in Himself, but also in the term of His creative act, God is always Creator, always Lord, since *always* means throughout all time. This, moreover, includes an additional consequence, that only in an analogical sense can we speak of God existing *before* creation. That there could be no *before* in time when time was not, is abundantly clear. That there was in the same sense no *before*

[39] Conf., xi, 15, 16; De Civit Dei., xii, 6.

in eternity, which has neither past nor future is no less clear.[40]

Though St. Augustine prefers to interpret, "the beginning" in which God created heaven and earth as the Eternal Word, "the beginning, who also speak to you," [41] yet he does not exclude the sense that God created all things in the beginning of time, which, as a fact, is the foundation of his literal interpretation of Genesis. Before that beginning there was no time. In that beginning God created time simultaneously with creatures. Though St. Paul speaks of the hope of eternal life promised by God before the eternal times [42]—thus St. Augustine translates what our English version renders in a sense the Holy Doctor would not deem foreign, "the times of the world"—yet what, the Saint asks, could be before those times? [43] So "God made all time simultaneously with all temporal creatures, which visible creatures are signified by heaven and earth." [44]

Because *always* means through all time, it follows that time was, as we have said, always, and God is always Creator and Lord. In this sense time, created simultaneously with material things, is, as in the text cited from St. Paul, called eternal. But it is not eternal as God is eternal. We do not term the world coeternal with God,[45] because this world is not of that eternity of which God is. God, indeed, made the world, and with it, time. Yet God is before time, since He is the Creator of

[40] Enarr. in Ps., ii, 6.   [43] De Gen. Cont. Manich., i, 3.
[41] John, viii, 25.   [44] *Ibid.*, ii, 4.
[42] Tit., i, 2.   [45] De Civ. Dei., xii, 15.

time. Similarly all things God made are very good, yet not in the sense that God is good, for He did not make them of His own substance, but of nothing.[46] "Hence Thou didst not create anything in no time because Thou didst create time itself. And no times are coeternal with Thee, because Thou remainest, and they, if they remained, would not be time."[47]

Speaking exactly, then, we do not say the world was created in time but with time. "For if the Sacred Writings, true in the highest sense, say, 'in the beginning God made heaven and earth,' so that He be understood to have made nothing before, beyond all doubt the world was not made in time, but with time."[48] "Nor will I suffer the questions of men who ask: 'What was God doing before He made heaven and earth; and what put it into the mind of Him who before had never made anything to make something?' Grant them, O Lord, to understand that *never* cannot be said where time is not. To say 'God never made,' what else is it than to say that He acted in no time? Let them see that without the creature, there cannot be time, and leave off talking nonsense."[49]

What then was that time created with the world? It was not time, but the beginning of time; as unity is not number, but the beginning of number. It was, as St. Augustine says profoundly, for we are at the origins of things, "the roots of times." "In the earth, indeed, as in the

[46] De Gen. Cont. Manich., i, 4. [47] Conf., xi, 17.
[48] De Civ. Dei., xi, 6; Serm., i, 5, init.
[49] Conf., xi, 40.

roots, so to speak, of times, these things had already been made, which were to exist in the course of times." [50]   The expression rests upon his ever present appreciation of the intimate necessary relation between existing creatures and existing time, between creation and primordial time. He saw the creature in its seminal reasons created in the roots of times; the creature tending to its existence in its own moment of time; the creature existing in its kind in the progressive course of time.   In other words, he saw the purely passive potency of matter determined simultaneously with its creation in that first instant, the roots of times; he saw that passive potency waiting for the appointed moment in time when it was to respond to the creative word, and, by creation, to become first of its kind; he saw the creature, so existing, continuing its existence, propagating by generation its species during time.

This analysis of the Saint's doctrine is confirmed remarkably by what at first sight seems almost a blunder on his part.   Though speaking of the first moment of creation, he calls it, not the root of time, but the roots of times.   Why the plural in place of the obvious singular?   How is a single instant, in itself indivisible, to be conceived as a multitude of roots; and why should it give origin to "times," not to "time"?   A little thought in harmony with that great mind, or rather carried on by its inevitable influence, gives the answer.   St. Augustine is not viewing by an abstraction time in itself, but in its concrete reality,

[50] Lib., v, 11.

in its inseparability from creatures, in which each
individual creature has its own individual time,
having its own appointed place in the universal
time of this coexisting order of creation. This
adequate concept necessitated the parallel view of
matter and time, viz.: prime matter in the initial
determination of its passivity, the seminal reasons,
not only collectively, but distributively also, of
all creatures that are to be: the first instant
of time created simultaneously with prime matter
so determined, containing in itself the root of
the time of every creature that is to exist, as
prime matter contains its seminal reason. From
these two in obedience to the creative word, will
come into existence at the appointed moment
in the course of time, the first of every kind by
creation; nor was it in the mind of St. Augustine
to conceive any intermediary agent. Indeed to
him such an agent would have appeared worse
than superfluous. He would have held it to be
impertinent. To one who understands this, the
argument of the Evolutionist which he urges so
confidently, drawing it from the armory of scholastic philosophy, in the principle, that beings are
not to be multiplied without necessity, is noise only
and nothing else. St. Augustine and those who
follow him reverently reduce the agents in creation to the lowest possible number. The Evolutionist multiplies them without necessity and without limit.

Nor can he help himself by calling attention to
the fact that St. Augustine does not here speak
of things hereafter to exist, as having been made

in the determination of prime matter, but in the
earth itself; that is to say, in the earth with its
elementary forms, their active potency, natural
forces, etc. None knew better than St. Augustine
what he explains clearly, that prime matter can
exist only under some form, and that every form
must have its active potency. But he allowed
those forms no formal evolutionary efficiency.
Having admitted the fact, he goes on to speak of
prime matter in its relation to what is to be, without any further reference to its form for the
moment. Indeed, any other course would be in-
explicable in one who, as we have seen, cannot
conceive the vital energy of the seed to reproduce
its own kind without the primordial determination
of prime matter to the receptivity, both in the
seed and in the coming plant, of this particular
form, by Him who created all things in their seminal reasons.

Here we find the very root of the difference
between the Evolutionist and St. Augustine. The
former, habituated to experimental science, assumes as something so certain as to need no proof,
that seminal reasons must be activities introduced
into matter working out to the orderly differentiation of species. The latter, penetrating beyond the
ordinary power of man, into the creative act,
places them as the necessary determination of
potency negatively universal, to those forms
which, thus created in the roots of time, should
each in the course of time become the term at
its appointed time of the one creative act.

This leads to the consideration of eternity and its relations with time.

St. Augustine tells us that eternity is before time, yet is not closed by time. " 'Our Lord's name,' says David, 'continues before the sun.' By 'the sun' are signified times. Therefore, His name continues for eternity. For eternity goes before times, yet is not closed by time." [51] There is then no passing from eternity to time, as if in the day of creation, in the roots of times, was the boundary between the two. This is fundamental. It seems obvious, yet it has to be insisted on; since, though admitted in word, it is ignored most frequently in discussion. Eternity and time are of different orders. They differ absolutely, as the Creator's eternal immutability differs from the transient mutability of creatures. Commenting on the words: "Thy years shall not fail," [52] St. Augustine asks: "What years are those that do not fail, if not the years which stand? If then the years stand, these and those are but one year, and this but one day; since this day has neither dawn nor dusk, and begins not from yesterday, nor is closed by tomorrow, but stands. You call that day what you will. If you wish, it is years. If you wish, it is day. Whatever be your thought, it stands nevertheless." [53]

Hence eternity is above the passage of time, yet it includes all time. Thus, as we have seen, God is, with reference to time, always the Creator,

[51] Enarr. in Ps., lxxi, 19.
[52] Ps., ci, 28.
[53] Enarr. in Ps., cxxi, 6.

yet the creature though bound up essentially in time, is not coeternal with Him.[54] For the creature there is past and future separated from the present by varying duration. The former, because past, has ceased to be; the latter, because future, is yet to be. For the creature there is indeed a present. Yet this is not an enduring present, else it would be eternity; but a changing present, going continually into the past. But how can that be said to be, of which the cause that it is, is that it shall not be?[55] St. Augustine does not pretend to exhaust the mystery. But he notes, nevertheless, that the past exists somehow in the memory, and that the future by expectation is seen conjecturally in its causes; while the present in continual progression divides the two. Thus is it with the creature. But not thus are past and future ever present to the Creator. This he illustrates very beautifully. One about to sing a song he knows perfectly, has the whole before him in expectation. When he begins, the action too begins to be stretched in opposite directions. What has been sung passes into the memory: what is yet to be sung extends into the future by expectation. Attention is fixed on the present, through which what was future is ever moving to become past. "But not thus dost Thou, Creator of the universe of souls and bodies, know things future and past. Far more wonderful, far more secret is Thy knowledge. Nor does anything happen to Thee, unchangeably eternal, that is, the truly eter-

[54] De Civ. Dei., xii, 15, 2.
[55] Conf., xi, 17.

nal creator of minds, as to one singing things
known, or to one hearing a familiar song, whose
affection varies and whose sense is drawn apart
by the expectation of future sounds and the re-
membrance of those past. As therefore, Thou
knowest in the beginning heaven and earth without any variety of Thy knowledge, so in the beginning didst Thou make heaven and earth
without any division of Thine action. Let him
who understands confess to Thee; and let him
who understands not confess to Thee." [56] Hence,
as in one simple, indivisible act God created all
things as they were to exist each in its own time,
so by one simple, unchangeable act did He know
them in all their mutual relations of time.

The same doctrine we find in *De Genesi ad Litteram,* though in different terms. "God, therefore, in His unchangeable eternity created simultaneously all things whence times were to flow,
and places were to be filled, and ages were to
revolve by the movement of things in time and
place. In them He created some spiritual, some
corporeal, forming matter, which not another, nor
no one, but absolutely He Himself instituted,
unformed yet formable, so that it preceded its
formation, not in time, but in origin. Over the
corporeal creatures He put the spiritual, such that
could be changed through times only, while the
corporeal could be changed through times and
places. Thus the soul is moved through times re-
membering what it had forgotten, learning what
it never knew, wishing what before it did not wish.

[56] Conf., xi, 38, 41.

But the body is moved through places from the earth to the sky, from east to west, and in other such manner. Whatever is moved through place must be moved also through time; but not everything that is moved through time is necessarily moved also through place. As, therefore, the substance moved through time only, precedes that moved through time and place, so is it preceded by that which is moved through neither time nor place. Wherefore, as the created spirit moved through time only, moves the body through time and place, so the Creator Spirit, moved neither through time nor place, moves the spirit through time. But the created spirit moves itself through time, and the body through time and place: the Creator Spirit moves itself without time or place; it moves the created spirit through time without place; it moves the body through time and place." [57] That is to say: God, immutable in His eternal present, is the immediate prime mover of creatures in all their vicissitudes of times and places. "Unless one believes that the substance of God is moved neither through time nor place, he does not yet believe God to be perfectly unchangeable." [58] But this calls for the necessary consequence that all time, past, present, future, is immediately subject to the eternal simplicity of the divine present. Thus, "All things were known to the Lord before they were created; so also after they were perfected, he beholdeth all things." [59]

[57] Lib., viii, 39.
[58] *Ibid.*, 43.
[59] Ecclus., xxiii, 29.

Hence the doctrine is clear. God, having decreed this existing order of creation created all things simultaneously in the beginning by His simple creative word. He created in the roots of time prime matter—under what elementary form or forms, is of no consequence—with its universal passive potency determined to those creatures only that were to exist in the course of time; and this is the creation of all things in their seminal reasons. This determination of passive potency was in no way the imposition of a form, but merely the adaptation of the universal *negative* potentiality of prime matter having no definite ordination to any form, to those it was actually to receive. Wherefore, seminal reasons were, with regard to the first members of any species originating by creation, purely passive. They were the primordial ordination of matter to respond to the creative word *in* time according to each creature's appointed time. In this response no intermediary agent intervened. To the creative word spoken in the simple unchangeable present of eternity, the response of each seminal reason was immediate, instantaneous, however widely separated it was from others in time; because the moment in time for the existence of each was immediately subject to the eternal, immutable present. Make the course of time as long as you please. Separate the beginnings of species by what duration you like. There can be no interval between the word spoken in eternity and its effect in time. St. Augustine leaves no room in his doctrine, as he proposes it, for Evolution.

His formula is as close as it is complete. The creative word, immutable, eternal, spoken once eternally in the eternal present of God, producing its effects, therefore, immediately in time yet according to the mutability of time, created simultaneously in the roots of times in their seminal reasons, that is, in the determination of the passive potentiality of matter to them alone among things abstractly possible, all creatures that were to exist, as they were to exist, each in its own time. We must observe that this "simultaneously" does not mean simultaneously with the creative act only, but also simultaneously among themselves. Whatever happens in time is simultaneous with the ever present moment of eternity. The simultaneous creation of creatures in the roots of times is opposed to the successive terminations of the creative act in the course of time, whereby each creature begins its own existence in its own time.

Having set forth the Holy Doctor's teaching in its genuine sense, we shall now proceed to show that it cannot admit Evolution as its complement, or consequence, or as any development whatsoever. If this be so, it is far from being fundamentally evolutionary. In the meantime, let us fix this firmly in our mind. Whether there be question of immediate creation without generation, or of mediate creation by way of generation, inasmuch as it is creation the seminal reason is always nothing else than the determination of the pure passive potentiality of matter to its future form. With this form the seminal reason is identified as its determinant. When existence begins, the determination is actuated. The seminal reason, hitherto purely passive, merged into the existing form becomes active, the seed of the activity of the seed generating similar forms. But of these it is *never the seminal reason,* the fundamental assumption of Evolutionists. It is a secondary cause, or better, perhaps, an intermediate agent in the actuating of the seminal reason of each, in which each was created when God spake and all things were created simultaneously in the beginning.

CHAPTER VI

THE ARGUMENT TO BE PROPOSED

THE preliminaries are over, and we have reached the very matter at issue. Influenced by his predilections, the Catholic Evolutionist assumes that the problem in *De Genesi ad Litteram* is to reconcile the simultaneous creation of all things in the beginning of time with their appearance in their various species successively in the course of time. Hence for him there is question of physical processes. The seminal reasons are for him an active potency given to matter under its elementary forms, which by its constant activity evolves gradually all forms of life, reaching eventually the existing multitude and variety of vegetable and animal species. For the passage from inorganic force to vegetative life, and from this to sensitive life, some demand a special divine assistance. Others, apparently more logical, are content with the ordinary Divine concurrence. The Thomist accepts the Saint's declaration that the problem is to show the perfect harmony between the literal sense of the history of the six days, found in the first chapter of Genesis and summed up in the first three verses of the second, and the literal sense of the single day of creation in which God made every plant before it grew up, etc., as expressed in the fourth and fifth

44

verses of the second chapter. Hence, he sees that
the Holy Doctor is dealing with the idea of creation, holding it to be in God one single act of absolute simplicity; in creatures to consist *formally* in
the creation of matter with its passive potency determined primarily and directly to those creatures
which without antecedent seed were, in obedience
to the creative word, to come into existence in their
various kinds; while *adequately* it includes the
successive appearances of each in its kind at its
own appointed time. With the existence of these,
seed-producing and propagating their kind, creation, formally and adequately considered, ceased;
and the present order, termed by the Saint, that
of administration, began. In it by conservation
and concurrence God moves all things according
to their natural active potency to the supreme end
of creation. But this active potency is what it
is in each specific nature, by virtue of the origin
of each. Its root is in that first actuating by the
Word of God of the corresponding passive potency to which that same Word determined and
restricted elementary matter in the beginning,
with regard to every creature that was to be
throughout all time. Hence, the seminal reasons,
inasmuch as they are the determination of prime
matter to definite future forms, are fundamentally
the material cause of all active potency to the end
of time. Actuated by Him who imposed them, in
the form to which He ordained them, they became with it the principle of active potency in all
things that exist, that have existed, that are to

exist. The former operation is creation: "In the beginning God created heaven and earth." With the latter, says the Holy Doctor, administration begins and in its iteration is continued: "My Father worketh hitherto and I work." [60]

The Christian Evolutionist, occupied with processes and physical agents, assumes the same mental attitude in St. Augustine, and interprets accordingly whatever may be drawn to his theories. The Saint, on the contrary, is taken up with origins. What connects the universal passive potency of matter with this particular order, not that? Why is there this succession of living beings, not that, to which matter as matter was equally in potency? Consequently evolutionary interpretations are foreign to his true meaning. To confirm this we shall establish the following propositions.

I. St. Augustine knows in general only two proximate origins of actual life, viz.: origin without seed, or creation, for the first individuals of each species; and generation by means of seed, the natural method of the propagation and increase, each in its own kind, of the species that originated without seed. Hence, the evolution of the perfect from the imperfect, the higher from the lower, the many from the few, does not occur to him even as a possibility.

II. Between the creation of material life in its seminal reasons, and the actual existence of the first individuals of each species, St. Augustine

[60] Lib., iv, 23; v, 27, 40.

puts no intermediate activity. Hence, this actual existence is but the adequate complement of the creation in seminal reasons, and is effected immediately by the creative word.

III. The creatures that began to exist without seed, each in its own kind, are, according to St. Augustine, those with which we are familiar, definite in their species unchanged to the present day. Hence, for him, existing species are the result of immediate creation, not of a long-drawn evolution.

IV. St. Augustine knows only two divine activities regarding living creatures, creation and administration. The former terminates with the actual existence of the first individuals of each kind. The latter begins with these, and consists principally in the conserving of them in their kind and in the coöperating in their propagation and multiplication, each in its own kind. Evidently there is here no idea of evolution.

V. The seminal reasons are, with regard to creation, a determination of the universal passive potency of matter according to the definite order of creation decreed by God to exist. They leave that potency purely passive, adding to it only an obediential relation of receptivity to the creative word, with regard to the form of each individual to come into being by creation in its own time. That St. Augustine in the treatise *De Trinitate* treats them in some way as active principles, implies no contradiction, since there he is not discussing creation, but speaking of administration; and consequently views these seminal reasons not only as limitations of universal passive potency of matter, but also as in formed matter, active in the forms to which prime matter was by them determined.

CHAPTER VII

CREATION AND GENERATION

ST. AUGUSTINE understands in general only two proximate origins of life; one without seed, the direct effect of the word of God creating it in its seminal reasons on that day unknown to us; the other by means of seed in the days we know of time. Let us see a summary of his doctrine. "What first was created was day; for that should hold the first place in creation, which could know the creature by means of the Creator, not the Creator by the creature. In the second place, the firmament, whence the corporeal world begins. Thirdly, the sea and land specifically, and in the earth potentially, so to speak, the nature of herbs and trees. For so the earth at the word of God produced them before they had sprung up, receiving all the numbers of what during the course of time it should put forth according to their kind. Then after this habitation, as it were, of things was established, the luminaries and stars were created on the fourth day, so that the superior part of the world was first adorned with things visible which move within the world. On the fifth day the nature of water produced at God's command what originated in it, that is, all fishes and fowls, and these potentially in the numbers that should be put forth through the congruous move-

ments of times. On the sixth day in like manner
terrestrial animals, as the last from the last element
of the world, but potentially, whose numbers time
would afterwards unfold visibly. This whole
order of the ordered creature that day knew: and
as that knowledge presented, in a way, the order
six times, the day, though only one, showed the
things that were made as six days." [61]

From this we gather that St. Augustine, having
in the preceding paragraph expressed prime matter as antecedent to the reception of the form, not
in time, but causally only, puts as the first of creatures, day. But this is not a physical day. It is
the double angelic cognition with its consequence
of praise, which he puts as the literal interpretation of the term in Genesis. He holds that the
firmament, the actual sea and land, the luminaries
of heaven and the stars were created in their individual existence, but all material life in its seminal reasons only. Thus creation was accomplished
formally according to determined numbers in the
roots of times, but not *adequately*. This called
for the actual existence of living beings, the first
individuals of each kind, to appear in the process
of time. In this way "God made heaven and
earth and every green thing of the field before
it was above the earth, and all grass of the field
before it sprang up." [62]

How, then, did these come to spring up? This
St. Augustine indicates in discussing what follows in the scripture text: "For God had not

[61] Lib., v, 14, 15.
[62] *Ibid.*, 16.

rained upon the earth and there was no man to till it." Both of these, rain and tillage, each in its own way, are necessary for the coming up of plants. Then both were lacking. Therefore, God made those first plants by the power of His word, without rain, without the work of man.[63] Elsewhere, however, he expresses his mind on this matter more at length and with greater clearness. "Where," he asks, "did God make these things in the day in which He made heaven and earth? Some say, in the Word of God, before they sprang up from the earth. But the scripture says distinctly, 'in the day.' Therefore, not in the Word of God, which is before the day. Were they in the earth itself causally and rationally as all things are in their seed before they, in a manner, evolve and unfold their growth and visible forms through the numbers of times? But the seeds we see are already on the earth, they have sprung up already. Or were they, not on the earth, but within it, and for this reason things were made before they sprang up, because they then sprang up when the seeds germinated, as we see happening now during the periods of time distributed to each, according to its kind. Were seeds therefore made then? Did the earth first produce seed? Not thus speaks the scripture: 'And the earth produced every herb giving seed according to its kind.' From these words it appears that the seed was from the plants; the plants, not from the seed, but from the earth." [64]

[63] *Ibid.*, 18.
[64] Lib., v, 9.

Here St. Augustine says clearly that the first production of plants from the earth was without seed. Afterwards each produced its own kind by means of seed. That he knows no other production is equally clear; for, making a formal enumeration of the different ways in which things exist, when he comes to actual existence he assigns only these two proximate origins. He is speaking of original sin, which, he says, must be found in Adam's sin when he was living his own individual life. "It would be sought in vain while he was still causally created in things created simultaneously, and was neither living his own proper life, nor was in parents so living. For in that first creation of the world, when God created all things together, man was made so that he should exist afterwards, the reason of man to be created, not the actuality of man created. But these (created things) are one way in the Word of God, where they are eternal, not made. They are otherwise in the elements of the world, where all things, made simultaneously, are future. They are otherwise in things which, created simultaneously according to their causes, are now created, not simultaneously, but each in its own time, amongst which Adam, now formed from the slime and animated with the breath of God, is as the grass that sprang up. They are otherwise in seeds, in which again are sought causes, as it were, primordial, drawn from things which existed according to causes which God created first, as the herb from the earth, the seed from the herb." [65]

Here, then, we have first, the definite assertion that with regard to creatures the seminal reason is but the reason of future existence, not an active cause such as Evolution demands. Second, that the actual existence of the first individuals of a kind is by creation. Adam was in the first creation of the world as the future man, the man to be created, while one living his own proper life is the man created. Third, man and plant alike, as they came from the earth without seed, came by creation. Fourth, the natural process of generation and of germination would be impossible without the influence of the seminal reasons, with which they are connected by the creation of the first individuals of the kind. And so he continues: "In all these the things which, already made, came forth with visible forms and natures from hidden and invisible reasons lying hid causally in the creature, received the modes and acts of their times, as the herb that sprang up over the earth, and man made a living soul, and such like, which, whether shrubs or animals, pertain to that operation of God which works up to the present. But these also carry with them invisibly themselves, as it were, again in a certain hidden power of generating, which they draw from those first beginnings of their causes wherein, before they rose up in the visible appearance of their kind, they were incorporated with the world created when the day was made." [66] Here, then,

[65] Lib., vi, 16, 17.   [66] Lib., vi, 17.

we have the two origins of things repeated. Things are created in their species as they are to exist; and this formally in their seminal reasons, adequately in their visible coming forth at their specific times. Things are generated, when those existing in time, in which God works constantly, reproduce themselves, that is their own specific nature, by seed. But this they can do only through that hidden power in the seed derived from the seminal reasons; and this derivation is possible only because these reasons, having determined the abstract universal potency of definite matter to particular species to terminate the creative act in the first members of such species, determined it in such a way, that each species was to be continued by generation. Wherefore, let us return for further confirmation to a passage already quoted in part: "The earth therefore is said to have then produced herb and tree causally, that is to have received the power of producing. In it, indeed, had been made already, as it were, so to speak, in the roots of times, the things that were to exist through the courses of times. For, to be sure, God afterwards planted Paradise in the east, and there brought forth from the ground every tree fair to behold and good for food. Yet we may not say that He then added to creatures anything He had not made before, which had, as it were, to be added afterwards to that perfection, whereby on the sixth day He finished completely all things very good. But all natures of shrubs and trees having been made already in the first creation, from which God rested to move

thenceforth and administer through courses of
times those same things He created, and from
which, when created, He rested, He then planted
not only Paradise, but also all things that now
spring forth. For who else creates these even
now but He who works ever until now. Nevertheless, these He creates now from the things that
now are: then, when absolutely they were not,
they were created by Him, when was made that
day, the creature, namely, spiritual and intellectual." [67]

So far, then, as creation in seminal reasons is
concerned, St. Augustine puts one essential difference between the planting of Paradise and the
daily growth of plants from seed. Their modes
of coming into existence differ. Paradise followed the completion of the six days' work and
the beginning of the rest of the seventh, as an
immediate creation, the adequate term of the
formal creation in seminal reasons, yet originating
without seed. Still, its actual external existence
added nothing to creation already perfect. This
can mean only that the formal creation in seminal
reasons and adequate creation in the creature
coming into existence without seed, constitute but
one indivisible effect of the creative word. The
creature's existence in time, in this time of the
world, not in that, is but the necessary verifica-
tion of the connection of its time with the time
of the world, both created in the roots of times. As
it comes into existence by virtue of the creative
act without adding to the perfect work of that
day known only to the Lord, so its own time that

[67] Lib., v, 11.

then begins with regard to the world and its time, as well as to other creatures and their times, adds nothing to the creation of all times in the roots of times. The creature falls into its place obedient to the divine Word. As regards primordial origins St. Augustine distinguishes not between the first creatures of their kind and those that follow them. All are created in their seminal reasons each with its own—it may be so expressed—seminal time, in the roots of times. But, for all that, he does not forget the fact that the former come into existence without seed, the latter generated by means of seed.[68]

Here the Evolutionist might perhaps object, that in his system every species may be said to come into existence without seed until the fixed species are reached, since no seed as such in the process of evolution corresponds absolutely to the species that by differentiation springs from it. This, however, would not be to interpret the mind of St. Augustine, but to read the Evolutionist's meaning into his text. Moreover, it would not rise above special pleading. Evolution is essentially a process of generation and of long continued generation, in which the generated is differentiated from the generator. St. Augustine's view is explicit. The first members of every species originate from the earth by virtue of the seminal reasons without seed. All others, with an apparent exception hereafter to be explained, come from those first members reproducing themselves,

*Lib., viii, 6.

that is, propagating their species exactly and invariably by means of seed, through the original determination of those reasons.

Another difficulty, more serious apparently, arises from the words of a previous quotation: "Though earth, therefore, was then said to have produced herb and tree causally, that is, to have received the power of producing." [69]   Whence, some say that the seminal reasons were indeed an active potency communicated to the earth enabling it to produce each species in its kind.   But such an idea comes rather from a predisposition to evolutionary theorizing, than from a patient study of the Saint's teaching, and cannot stand with this completely grasped.   The only active potency in the earth as such was that of the elementary forms, quite inadequate to the production of the varied life of the vegetative and sensitive creature.   Indeed, this was so obvious, that, though St. Augustine recognizes the existence of such forms, since prime matter could not exist uninformed, he nevertheless ignores them in discussing the seminal reasons as the term of the first creation, putting these, as we see, in prime matter as a pure passive potency.   This, he calls, the receiving by the earth of the power of producing, because he sees that in its concrete determination to creatures decreed to exist, the potency of matter received a reality.   Such reality, though actuating no passivity, yet, when taken in comparison with that potency, universal, undetermined,

[69] Lib., v, 11.

able to be anything, ordained to nothing, may well be called a power; since by its virtue prime matter will, without any intermediary agent, respond in living creatures to the creative word, under the proximate activity of which it lies until in the processes of time its time shall come for formal determination to the future being to pass into the formal actuation of the being present and existing. On the other hand, if, according to St. Augustine, seminal reasons are active with regard to the origin of the first members of each species, they must be, as Evolutionists wish, functions of generation; and so the Saint's distinction, origin without seed by the word of God, origin by seed in propagation, vanishes.

All this being premised, we must explain the words in question: "The earth produced herb and tree causally, that is, it received the power of producing." In the first place, then, the prime matter determined to the form of this particular herb or tree was its material cause, and had causality as such. Secondly, it existed with that determination under some elementary form, though what that elementary form was had nothing to do with the seminal reason. Hence, matter as such is rightly called the earth, and the seminal reason is in the earth. Thirdly, in the actuation of the seminal reason in which matter received the form of herb or tree, it had to lose its elementary form; and so matter thus viewed could no longer be called the earth, nor looked on as remaining within the earth. To lose its elementary form and to receive actually the higher forms of herb and tree supposes some disposition of the matter. This proximate disposition comes immediately from the creative act. Remotely, mediately and instrumentally, however, the natural modifications of the earth, whereby it became a fit habitation for plants and trees had their effect. Thus are explained amply the words quoted.

Nevertheless, we must be allowed to repeat what we can never insist upon sufficiently, which, if understood, must remove the last remains of doubt. Though St. Augustine regards prime matter determined to its particular species as still purely passive, he sees that it is no longer such negatively, but positively. The determination is a reality giving matter a real power, not of *acting* but of *receiving*, without which it could not be under this form rather than that; and so a definite creation distinguished from others would be impossible. This power of receiving forms that must rise from the earth at their appointed time, imposed on certain matter and not on other, may well be predicated reductively of the whole earth, as a power of producing.

. It must be borne in mind that we are dealing with something unique, beyond all experience, increasing in mystery the more deeply it is penetrated. Man's language, therefore, is inadequate to its perfect expression; as none knew better than St. Augustine, who felt the mystery, as we can never hope to. We must look, therefore, for analogies of expression, rather than for a con-

stancy in univocal terms. It would be a grave
mistake to assume that in our understanding of
St. Augustine's doctrine, we suppose a first merely
voluntary determination of the passive potency
of matter to all its actually future functions,
and a second actuating determination at the moment of each, to give the actual existence. There
is but one determination coming from the creative
word, spoken once, eternally, incapable of repetition. Indifferent, not negatively but positively,
to the beginnings of times and to all times, it is
equally efficacious in determining potency, giving it
power to receive this form, and in actuating it with
the form when the moment of existence comes. In
both cases its effect must be real. Indeed, referred
to the act, they are not distinguishable except by
a notional distinction, whatever be the real distinction in the term. But whatever differences of
times we see in the term, every element of it is
connected immediately with the creative act.
Matter receives its power of reception; this power
is actuated. Between the two there is no intermediate activity, nothing but the unfolding of its
time, until the moment comes that, coinciding with
the time of the world according to the divine decree, marks the passage of the creature into visible
existence from this potency of creation, which we
may, in a sense, accept as already actuated, so certain is its term.[70]

But this is to be dwelt on more fully in the next
chapter. Here we have said enough to show how

[70] De Gen. cont. Manich., i, 11; *cf. supra*, p. 19.

the inadequacy of our human language compelled
St. Augustine to use the expression; "the earth
received the power of producing" not, however,
univocally with our common mode of speech, but
analogically.

CHAPTER VIII

BETWEEN CREATION IN SEMINAL REASONS AND ACTUAL EXIST-ENCE, NO INTERMEDIATE ACTIVITY

FROM all that has been said, St. Augustine's doctrine is so clear as to the absence of all intermediate activity between creation in seminal reasons, and the appearance of the creature in time, that this chapter might seem superfluous. Nevertheless, so deep seated is the notion, antagonistic though it be to any adequate concept of creation, that unless multiplied evolutionary activities be allowed, it is necessary to admit successive creative acts, that this alone gives sufficient reason for the discussion on which we are entering, or, to be more exact, demands it.

In the first place, St. Augustine gives no hint of any such intermediate activities. Creatures are created in their seminal reasons, to be put forth from the ground, when their time shall have come, in their actual existence. The Word creating them covers all. However, a little study will show that we have on this point grasped indubitably his real mind. In a passage already quoted, he insists on the production of the first individuals of each species, not from seed but from the earth, establishing the fact on this, that God did not say:

"Let seeds germinate," but: "Let the earth germinate—and the earth brought forth." [71] "Let the earth germinate." This again seems to present a difficulty. Germination is certainly an exercise of active potency, which seems to contradict our explanation of St. Augustine, that in creation the seminal reasons do not exceed passive potency in their determination of the potency of prime matter. We might reply with perfect justice that analogies are not to be pushed too far. They are to be limited by what is otherwise certain; what is certain is not to be upset by them. The very idea of an analogy requires not only agreement on certain points but also disagreements on all others. The earth is not a seed; and one goes beyond his right in assuming contrary to what has been established, that St. Augustine would have the production of creatures from the earth follow in everything the process of seed germination. We can, however, do better. We can take their analogy and confirm from it the doctrine of our preceding chapter, that St. Augustine had not the least idea of granting any material force a formal share in the origin of the first creatures from the earth. A little farther on from the words just quoted he resumes his comparison. "Let us therefore consider the beauty of any tree in its trunk, branches, foliage, fruit. It certainly did not spring up suddenly in this outward appearance, but by the orderly process which we know. It grew from the root fixed by the first germ in the earth, whence all these things grew up in this

[71] Lib., v, 9.

distinct formation. But the germ came from the seed. In the seed all those things were primarily, not in their corporeal mass and size, but by causal force and potency. For that size is built up from the supply of earth and water, but in the little seed is that more wonderful and noble force, which is able to take water, mixed with earth as material, and change it into that kind of tree, its spreading boughs, its green mass of foliage, its abundance of fruit of particular form, and all these things in their most distinct order. For what comes from the tree, that is not drawn in a hidden manner from the seed? . . . But as all that in the process of times sprang up in the tree, were all together invisibly in the seed, so must we hold that the world, when God created all things simultaneously, held simultaneously all things which in it and with it were made, when day was made; not only heaven with sun, moon and stars, constant in their specific rotary motion, and earth and the abysses, that as it were, suffer irregular movements, and joined to heaven from below, give its second part to the world; but those things also which the water and earth produced causally and potentially before they sprang up through intervals of time, as they are now known to us in those works which God works to the present moment." [72]

To one who holds germination to be a process merely chemical, or who grants to chemical force an efficiency other than merely instrumental, this passage will appear conclusive, an assertion clear and emphatic that for the first production of crea-

[72] Lib., v, 44, 45.

tures in their species, active forces worked in the earth, as they do in the seed for the species' propagation. Yet such an interpretation would be but the futile reading of one's own mind into St. Augustine, instead of the discovering of his by patient labor.

We have seen already how the Saint viewed the yearly round of sowing, growth, maturity, and harvest, as no less wonderful than the changing of water into wine; and found the mystery of generation and birth as awe-inspiring as the resurrection of the dead. The occurrences of every day, losing their wonder as they become common, demonstrated the divine power to him, as clearly and surely as the feeding of the five thousand, the water changed into wine, the raising of Lazarus.[73] He never wearies of repeating with St. Paul: "Neither he that planteth is anything, nor he that watereth, but God that giveth the increase.[74]

This he repeats in the treatise before us. "Who does not know that water, mixed with earth, when it reaches the roots of the vine is drawn into the sustenance of that wood, and receives in it the quality by which it becomes the gradually appearing grape cluster, and that in this it becomes wine, and maturing grows sweet, which, when pressed out, ferments and comes, when made stable by some age, to serve more profitably and pleasantly as drink? Did the Lord on this account seek wood and earth and these intervals of time, when

[73] *Supra*, p. 24.
[74] I Cor., iii, 7.

by a wonderful short cut He changed water into wine, and such wine as won the praise of the feaster already sated? Did the Creator of time need the aid of time? . . . Nor, when done, was this done contrary to nature, except as regards us who know the course of nature differently. But not for God to whom nature is this which He has made." [75]

To God, then, it is equally according to nature to create directly, to create indirectly by generation, or to produce miraculously. All three depend on His almighty power, not upon nature or natural forces, which of themselves are as unequal to the task of producing wine through the long process of assimilation, growth, maturity, fermentation, as to the taking of the miraculous short cut of Cana, or to the producing of the first vine from the earth.

Of this omnipotence the causal reasons are in every case the immediate effect. To them is to be referred, what is their proper function, every determination of prime matter in the existing individual, as it shall come into existence, to the end of time. This is stated distinctly in a passage lately quoted: "But as whatever things sprang up in the tree in the process of time were all together invisibly in the seed, so must we hold that, when God created all things simultaneously, the world held simultaneously all things made in it and with it, . . . those also which the water and the earth produced causally and potentially before they sprang up through the intervals of time, as they

[75] Lib., iii, 24.

are now known to us in the works which God works to the present moment." [76]

That seminal reasons as positive determinants of passive potency were, as was to be expected, indifferent as to the active force hereafter to actuate them, whether the word of God directly without intervening agency of any kind in the first creatures of each species, or the word of God indirectly by means of natural generation, is not merely deducible from St. Augustine's words. He asserts it distinctly. "Were causal reasons established . . . to pass through definite terms according as we see all things coming into life from shrub or animal, . . . or were they to be given straightway their fullness of form, as Adam, believed to have so received it without any progression of youth? Why not in both ways, so that from them might be in the future what the Creator had decreed? Should we say the former mode, miracles contrary to the usual course of nature would appear contrary to them? If the latter, greater absurdity follows, that in passing through their periods of time, the very forms appearing daily contradict the primary causal reasons of things coming into life. We conclude, then, that they are created adaptable to either way; to this, by which temporal things pass most commonly to their perfect state, or to that, by which rare things and wonderful are done, as it shall have pleased God to do what the time demands," [77] In themselves the seminal reasons, regarding primordial origins, natural genera-

[76] Lib., v, 45.
[77] Lib., vi, 25.

tion, miracles indifferently, are but passive determinations of passive potency to be actuated according to the requirements of each.

In two modes, therefore, are seminal reasons
brought to existence, immediately, without any
antecedent process of generation, and mediately,
the affects of that process. To the first mode belong the creation of the first individuals of the
species and the miracle. Both are instantaneous,
excluding that progression through determined
times from the first elements of being to its full
perfection, the essential note of generation. The
miracle differs from the first production of creatures in this, that occurring in the existing order,
it occurs out of the natural course, while the latter,
of its nature antecedent to the natural course, begins it according to God's decree. Hence, St.
Augustine excludes the first individuals of each
species from those very processes of nature in
which Evolution consists.

We must not omit to note that, as regards the
natural process of production by generation, St.
Augustine does not *exclude* the lower agencies that
have in it their instrumental place. He takes them
for granted, and then passes them by, as having no
formal efficiency in the generating of the effect.
For this he fixes his attention upon the mysterious
vital activity, that assimilating the elements,
changes them into flowers and fruit, thus perfecting the seed in which the parent plant is to live
again. This vital principle, the energy of the
specific form, what is it ultimately in each individual but its seminal reason, created in passive po-

tency when God created all things simultaneously,
now after successive generations come into existence in its own time obedient to the creative word?
Thus the backward glance sees individual seminal
reasons specifically the same, receiving existence
according to their times, until that is reached which
ungenerated began the species and the specific
time, the immediate adequate effect of the one all-
embracing creative word. And so we read: "God
created all things so that, what we now see, creatures moved in intervals of time to accomplish each
what belongs to its own kind, was to come from
those implanted reasons, which God scattered
seminally, as it were, in the instant of creating,
when 'He spake, and they were made; commanded, and they were created.' " [78]

But the expressions, "in its own time," "in their
own times," remind us that for all alike, whether
generated from seed or coming to exist without
seed, St. Augustine recognizes one process, one
progression, to which he attaches such importance,
as to almost weary the reader with its iteration.
Let us see this in a summary of the six days, already quoted,[79] in which, on reaching the creation
of living beings, he insists in each case on their
creation with numbers and times. "Numbers"
means, as we shall see more at length, that the
seminal reasons were fixed, each kind in its number, determined by the decree of creation, according to the exemplary ideas in God and the corresponding creatures that were to exist; so that each

[78] Ps., xxxii, 9; Lib., iv, 51.
[79] Supra, p. 49.

individual creature coming into existence to the
end of time is the adequate term of its own
seminal reason, created in the beginning of time;
while the last so to come into existence will exhaust the sum total of the seminal reasons in which
creation terminated formally, when all things were
created simultaneously,[80] Let us, therefore, come
to what concerns us here.   What are these processes of time?   Why does St. Augustine insist on
them so earnestly?

With this question in view we premised the
chapter on St. Augustine's concept of time, of
eternity and of their relations.   Time is neither the
being consisting of matter and form, as it moves
continuously from the beginning to the end of its
existence, nor is it the motion considered in itself.
Yet, it is necessarily connected with such creatures;
so that a material being tending to corruption without time, or time without such a being would be
inconceivable.   Hence, St. Augustine teaches that
time is a creature, created simultaneously with the
material creation when God created all things together.   What the absolute measure of the time of
the universe is, a question plunging the enquirer
into mysteries, does not concern us.   The relative
measure for all creatures is evident, namely, the
secular movements of sun, moon and stars in
heaven.[81]   These, by their constancy, and by their
duration lasting as long as time shall endure, constitute a measure for the briefer durations of beings that come into existence and depart.   Yet

[80] Lib., iv, 7-12.
[81] Lib., ii, 28.

these again in their own movement, whether this
be referred to the movement of the heavens, or
taken independently as a part of their own duration, measure also the movements of subordinate
beings that come into their existence. Each material being, therefore, has its own time created
with it, in which it begins its existence at a determined moment in the successive revolutions of
the heavens, and continues simultaneously with
them. By this can be measured the movements of
those beings that minister to its existence until the
determined moment of the world is reached in
which the individual's time ceases, and the creature of a day vanishes from time.

Time, thus considered, has its origin, as have
the things of time. These, created in their seminal
reasons on the day when God made all things simultaneously, were not actual existences; and so that
day was not time. As things were then but
seminal reasons, so in that day were the roots of
their times ready to begin actually, when things
should actually begin to exist. An actually existing material being without time, and time without
an actually existing material being, are inconceivable. If, therefore, the seminal reason is to be
brought to actual existence, this must be in its own
time, to last during its own time and to cease by
natural corruption when its own time shall have
been fulfilled. This time is measured by that of
the world. As the world's time rolls on, and the
moment comes which coincides with the destined
first moment of the creature's time, the seminal
reason is actuated. In obedience to the all-suffi-

cient creative word, once uttered, never repeated, real existence begins. Thus, St. Augustime connects the first existences with the evolution, if you will, *of times,* but never with a process of evolution *in time.* Yet this he must have done, had he conceived seminal reasons as active forces evolving the species to be. In his mind the seminal reason, that unique determination of passive potency as we explained it, always immediately subject to the creative word, needed only the occurring in time of the first instant of its own appointed time to spring into being.

This St. Augustine explains very beautifully. "Were those first works of God not perfect in their way, what was lacking to their perfection would doubtless be added to them afterwards, so that each furnishing, as it were, a half, as if parts of a whole to be completed by their union, a certain perfection of the universe would arise from their conjunction. Again, if those first works were perfect, just as they are perfected when, each in its own time, they are brought forth into visible forms and activities, it is certain either that nothing would be made of them afterwards through times, or that this would be made which God did not cease to work from these things which now arise each in its own time. But now, those very things which, in the beginning when He made the world, God created simultaneously to be evolved by succeeding times, are in a certain way already consummated and in a certain way begun. They are consummated, because in their own natures, by which they ac-

complish the courses of their own times, they
have nothing that has not been made causally in
these. They are begun, because they were, so to
speak, seeds of future things, to be brought forth
out of their concealment visibly in suitable places
through the extent of ages. Wherefore we can
gather this from the scripture, for it calls them
both consummated and begun: 'So the heavens
and the earth were finished (*consummata sunt*)
and God ended (*consummavit*) His work.' On
the other hand, unless they were begun, the passage would not continue: 'God rested from all
his works which he began to make' ἤρξατο ποιεῖν
(*Septuagint*). If, therefore, one should ask, how
God consummated and how did He begin, it is
clear, from what we have said, that the things
He consummated were not other than what He
began. We understand that God assuredly consummated them when, creating all things simultaneously, He created all so perfectly that nothing remained to be created in the order of times,
which had not here already been created by Him
in the order of causes. We understand Him to
have begun, so that what He had predetermined
in causes, He would afterwards fulfil in effect." [82]

Here St. Augustine considers three conditions
of creatures, their creation in seminal reasons,
their coming into existence, each in its own time
and place, and God's dealings with them after
they come into existence. He then argues: Unless they were perfect in their seminal reasons,
further creation would have been needed to bring

[82] Lib., vi, 18, 19.

them into existence, contrary to the word of scripture, "God made all things simultaneously." Unless, on the other hand, they were in their seminal reasons incomplete as regards their actual existence, it would follow, either that from those reasons nothing was made through the processes of time, which again would contradict the scripture, or else that, having received actual existence in the day when all were created simultaneously, God's action in them would be confined to the work of administration, that is, to their propagation by generation and seed. This, too, would contradict the scripture. Hence creation, beginning and accomplished formally in the seminal reasons, is perfected and accomplished adequately in the bringing of creatures into existence, as one operation of the one creative word.

It may be said that this passage can be harmonized perfectly with Evolution. But we must repeat again, that the question is not, what meaning Evolutionists can fit into the material words, but what sense they receive from their author. Moreover, a little closer investigation shows that such a perfect harmonizing would be a task anything but easy. St. Augustine, as we have seen, and shall see again, divides God's operations in creatures into creation and administration. Creation in its usual strict acceptation terminates with the existence of the first creatures of their kind, coming into existence ungenerated in obedience to the divine command. With their existence in their specific kind begins administration of creatures propagating their kind by generation.

Evolutionists would reach the species, that are for St. Augustine the term of direct creation, through a long process of successive generations, which, had it entered into his mind, the Saint would have put necessarily under administration. However, we can gather from his own testimony what would have been his idea in the case. He says distinctly that Adam's body was not created differently from those of other creatures;[83] that Adam was created according to his seminal reasons;[84] that he came into existence, not by generation, but suddenly in the bloom of manly vigor,[85] and that he was created from the slime.[86] "Wherefore, 'God formed man from the dust and slime of the earth, and breathed into his face the breath of life, and man became a living soul.' Not then predestined; for this was done before the world in the fore-knowledge of God. Not then causally, either begun in consummation, or consummated in inception; for this was accomplished from the beginning of time in the primordial reason, when all things were created together. But created in his own time visibly in his body, invisibly in his soul, consisting of body and soul." [87] A passage so clear and so perfectly to the point calls for no comment.

Could any doubt remain, St. Augustine's summary of his doctrine, in which he confessed that, after all his prayer, all his investigation, all his discussion, he finds himself face to face with

[83] Lib., vi, 20, 21, 22, 30.
[84] Ibid., 26, 29.   [85] Ibid., 23, 35.
[86] Ibid., 17, 26, 30.   [87] Ibid., 19.

mystery, should surely remove it. Let us therefore hear it in his own words. "If I shall say
man was not in that first creation of things in
which God created all things simultaneously, not
only not as a man of perfect age, but not as an
infant, nor even as an unborn babe in its mother's
womb, nay, not even as the seed that generates
man, he (*i.e.*, one not grasping his meaning) will
think him to have been absolutely non-existent.
Let such a one turn again to the Scriptures. He
will find man, male and female, made on the sixth
day to the image of God. Again let him see
when woman was made. He will find it outside
the six days; for she was made when God further * formed from the earth the beast and the
fowls. But then (*i.e.*, on the sixth day) man
was made, male and female. Therefore, then
and afterwards. Not, then and not afterwards;
nor, afterwards and not then; nor, different afterwards; but the very same, in one way, then, in
another afterwards. He will ask me how. I will
reply, *afterwards* visibly, as man constituted in
his exterior form is known to us; not, however,
by generation from his parents, but he from the
slime, she from his rib. He will ask me how,
then. I will answer, invisibly, potentially, causally, as future things as yet unmade are made.
Perhaps he will not understand. . . . What,
therefore, can I do but admonish him to believe
God's scripture, that man was made then, when

* Further. *Adhuc.* Greek ετι. From the Septuagint
St. Augustine explains it as referring visible production
to seminal reasons. (Lib., ix, 1.)

God made all things together, and also then, when no longer simultaneously, but creating each in its proper time, He formed man from the slime of the earth, and woman from his bone? For neither was made thus on that sixth day; nor, for all that, does the scripture permit one to understand that on the sixth day they were not made." [88]

Here, then, St. Augustine, while confessing the mystery, asserts, with a most distinct particularity, the immediate creation of Adam from dust and slime, of Eve from Adam's bone, and the identity of this creation with that primordial creation in seminal reasons. It would, therefore, be more than rash to split open his doctrine, in spite of the simultaneous iteration of the fact and acknowledgment of the mystery, so wide as to be able to introduce between the adequate effect of creation, and the primordial seminal cause, a long evolution of similar causes growing in perfection with each successive generation.

If, notwithstanding all that has been brought to bear, one still imagines that, to avoid multiplying successive creations, the doctrine of St. Augustine demands such an evolutionary explanation, let him recall the Saint's teaching on time and eternity. Eternity, simple, indivisible, stationary, without past or future, outside the concept of time, dominates time that is necessarily in real relation with it. For though it be unchangeable in itself, all the moments of time coincide with it positively, and it with them. Nor

[88] Lib., vi, 10.

does this relation touch one instant of time more
than another, but regards all equally.   The creative word, spoken in eternity can have its immediate effect, not only in *any*, but in *every* instant of time.   This effect beginning in the roots
of times, completed adequately in the course of
time after any imaginable duration, if the effect
of creation in the strict sense, is but one immediate effect of one immediate cause, with which
it is no more closely connected in its beginning
than in its term.   Nor is there any other limit
to such effects than the decree of creation.   One
who grasps the idea in St. Augustine's profound
formula: "God truly eternal, truly immortal and
unchangeable, unmoved either through time or
place, moves his creatures both in time and in
place," [89] can hardly hear without impatience of
the need of saving the Saint from multiplying creations and entities.

[89] Lib., viii, 40.

CHAPTER IX

THE CREATURES THAT BEGAN TO EXIST WITHOUT SEED ARE, AC-CORDING TO ST. AUGUSTINE, THOSE WITH WHICH WE ARE FAMILIAR

HAD we closed our work with the last chap-ter, we should have been justified in doing so. To anyone willing and ready to rise above preconceived notions, accepted, too often, from other persons, rather than from a careful study of the Saint's own words, his mind must now be clear. Indeed, one thing alone, his constant dis-tinction between the vegetative and sensitive beings coming into existence without seed, and their successors in the same kind as themselves produced by generation, is irreconcilable with any accepted theory of Evolution. Even the most mitigated Darwinism supposes a few primary, determinable, ancestral types, corresponding to the generic notion, rather than to the specific, to be so differentiated by successive generations un-der various conditions as not to reproduce their own kind, but to produce numbers of different kinds.

But Evolution is not willing to stop there. If fixed species in all their variety can originate by differentiation from such a primitive determinable type, why not this from something still more determinable, still less determined and differentiated? Thus protoplasm is reached, and through it what is most elementary in matter and force as we conceive them, and to what behind these may be still more elementary, until finally one utterly universal, absolutely undifferentiated, be reached, to be the logical starting point of Evolution.

So Spencer conceived it; and if we follow the Catholic Evolutionist, in accepting, because others have done so, St. Augustine's seminal reasons as active forces created with prime matter under its most elementary form to work out in the long processes of time all the varied vegetative and sensitive life of our world, we must class this holy doctor with Spencer, rather than with Darwin. St. Augustine, than whom amongst mortal men none approached nearer the throne of the Eternal or sounded more deeply divine mysteries, to be associated even in a passing thought with the coryphæus of all modern agnosticism, pantheism, atheism! The bare suggestion is intolerable. Let us, then, proceed; and out of the Holy Doctor's abundance draw further proofs that between his teaching and Evolution lies a chasm impassable. We shall show, therefore, that according to St. Augustine, the creatures that began to exist without seed are those with which we are familiar, definite in their species. Hence, existing species are for him the result of immediate creation, not of a long drawn evolution.

But with prejudice goes a proneness to misconceive. Let us, then, recall the warning that St. Augustine's object is not to work out a theory of the origins of the existing world, but to put before us, as far as possible, the operation of the Creator as revealed in the literal sense of the holy scripture. He does not say, therefore, that these existing species are the adequate term of divine creation, so as to exclude others once existing and now no longer seen. Of itself the matter lies outside his field. Nevertheless, it comes in accidentally; and so, speaking, not as a naturalist, but as one pondering the mysteries of the roots of times and that day of creation, which is not as our days, he says: "This universal creature of God has many things we know not, either what are higher in the heavens than our sense can reach, or in regions of the world perhaps uninhabitable, or which lie hidden below us in the depth of the abyss, or in the secret recesses of the earth." [90] But such, he goes on to say, though unknown to us, were known, both as they are in God and as they are in themselves, to that day of evening and morning knowledge, to which all things are known. Had he, therefore, been acquainted, as we are, with the monsters of the primeval world, far from changing or even from modifying his doctrine, he would have found in the fact of creatures so vast, so wonderful, seen by no human eyes, revealed only to angelic intellects, not a suggestion of Evolution, but the confirmation of his ideas, proposed

[90] Lib., v, 36.

so profoundly and yet so modestly, of the six days of creation.

It must be equally clear that, in speaking of existing species, St. Augustine speaks of facts, namely that such species actually exist in the animals that surround us; and that he does not enter into the naturalist's questions, of whether this animal and that are distinct in species, or whether they are only permanent varieties of the same species. That there not only can be, but ought to be, such permanent varieties in determined species, is clear to the scholastic philosopher; and must have been equally clear to St. Augustine, who differs from the former, not essentially, but in mode rather, and terminology. The question, however, does not enter into the literal sense of Genesis, with which the Saint was occupied.

These two points being understood, let us come to the matter of the present chapter in the words of the Saint himself. He is discussing the literal sense of Genesis i, 24, 25, which he quotes according to the Septuagint: "And God said: Let the earth bring forth the living soul according to its kind, quadrupeds and reptiles and beasts of the earth, according to their kind, and cattle according to their kind. And so it was done. And God made the beasts of the earth according to their kind, and cattle according to their kind, and all reptiles of the earth according to their kind. And God saw that they were good." On this text he speaks as follows: "The kinds of animals the earth produced in the Word of God are manifest. But because by the term, cattle, or the term, beasts, all animals without reason are often understood, it is right to enquire what the scripture calls beasts in the strict sense, and what, cattle. There is no doubt that by creeping things or land reptiles it would have us understand all that creep, although they, too, can be called beasts. Again the term, beasts, is in common use to express lions, leopards, tigers, wolves, foxes, dogs also, and apes, and other such like animals. The name, cattle, is given more suitably to those which man uses, either to help him in his work, as oxen and horses, and similar animals, or wool-bearing animals, or those used for food, as sheep and swine. What then are quadrupeds? Although every animal we have mentioned, some few reptiles excepted, go on four feet, still, unless by this name some particular animals were to be understood, the scripture certainly would not name quadrupeds here, even though it omits them in the repetition. Red deer, fallow deer, wild asses, wild boars surely can not be put with lions among the beasts. They are like cattle; though, for all that, they are not domesticated by man. Are they meant by the term, quadrupeds, as though left to receive in a special sense a term common to all going on all four feet?" [91]

Here there is question of the work of the sixth day necessarily identified in St. Augustine's doctrine with the day unknown to us, wherein God made all things simultaneously. There is ques-

[91] Lib., iii, 16, 17.

tion, then, of seminal reasons identical with the animals we all know, in which they were to exist. "Not one thing then and another thing afterwards, but then and afterwards the same," [92] differing in this only, that what then was invisible was afterwards visible. The term of creation in seminal reasons, therefore, was not force or active potency in some elementary forms capable of evolving the higher creatures, but the very creatures themselves, as they were to come forth in their own times visibly from the earth. In discussing the passage to which we alluded towards the end of the preceding chapter: "God further formed from the earth all beasts of the field," [93] St. Augustine remarks: "If what has been considered and written in the earlier books is of any aid to the reader, we need not dwell on this, that God further formed from the earth all beasts of the field, more than briefly, to ask why the word, *further*. It is on account of the first creation of creatures consummated in six days, in which all things together were causally completed and begun, so that afterwards the causes might be carried out to their effects, as we have intimated to the best of our power." [94] Again, touching the same matter, he asks: "If, then, in consequence of no helper like to man being found among the cattle, and the beasts of the field, and the fowls of the air, God made him one from a rib of his side; and this, moreover, when He had further formed these same beasts from the earth

[92] Lib., vi, 10.   [93] Genes., ii, 19.
[94] Lib., ix, 1.

and had brought them to Adam, how are we to understand that this was done on the sixth day? . . . This, therefore, would not be said: 'and God further formed' etc., were it not that the earth had already produced all beasts of the field on the sixth day. Therefore, otherwise then, that is potentially and causally, as suited that work in which He created all simultaneously, from which He rested on the seventh day: but otherwise now, as we see those things which He creates through the spaces of time, as He works even to the present." [95] And so the Saint concludes that as Adam and Eve were both created in the seminal reasons of their bodies on the sixth day: "Male and female created He them," [96] and yet were brought into actual existence in these well known days of corporal light which are caused by the course of the sun, so also was it with the first creatures coming into existence without seed.[97]

The term, therefore, of the creative act was in the seminal reasons the creatures themselves that were to be brought out of the earth at their appointed time, not evolutionary forces in matter to evolve them. But they were in prime matter invisibly, causally, as that matter was determined to the reception of their forms at the time of each. Till that moment came, matter remained so determined passively under the immediate activity of the eternal creative act, itself unmoved yet moving its creatures in place and time. Then without any further activity, simply as the ade-

[95] Lib., vi, 7.   [96] Gen., i, 27.
[97] Lib., vi. 7.

quate term of creation, prime matter thus passively determined under the immediate action of the creative act, positively indifferent to all times and all places, received in their own times and places its substantial forms, and living creatures came forth visible and actual. Thus came into actual existence the first members of the species familiar to us today, according to every Evolutionist the result of many a transient form, according to St. Augustine without any intermediate activity of any agent. This is the clear teaching of St. Augustine, and in its profundity worthy of such a doctor.

CHAPTER X

CREATION AND ADMINISTRATION

ST. AUGUSTINE knows only two divine activities with regard to the life of creatures, creation and administration. The former terminates with the actual existence of the first individuals of each kind. The latter begins with these, consisting principally in the conserving of their existence and in the co-operating with their propagation and multiplication, each in its own kind. Here there is no room for Evolution. The conclusion is so evident, that, were it not for an apparent difficulty against the antecedent assertion, this chapter would be superfluous.

The difficulty is this. St. Augustine makes creation end with the work of the six days: "In the beginning God created heaven and earth;" and administration begin with the Sabbath of rest: "My Father worketh hitherto and I work." But the work of six days does not go beyond the creation of all things in their seminal reasons, simultaneously in the roots of times; this visible production from the earth occurred in the processes of times, and therefore comes under administration immediately, *i.e.*, conservation, propagation, and is only mediately referable to creation. Therefore, not only is there room for Evolution, but it is actually demanded. The major appears evident from several texts we have already quoted, in which this distinction is asserted. Thus: "In the first creation from which He rested on the seventh day, God effected creatures in one way. He effects their administration by which He works to the present moment, in a way quite different. Then He effected all together without any definite intervals of time; now through definite intervals of time." [98]   As regards the minor, it is certain that St. Augustine admitted as evident that the visible production of creatures occurred at intervals of time. Hence, the conclusion seems unavoidable.

In answer we might appeal to the consistent doctrine of the Saint, gathered from many sources and demonstrated to an absolute certainty. Should a conclusion inconsistent with it seem to flow from a passage here and there, the seeming contradiction must not invalidate the concordant exposition of the Saint's mind, but must be attributed to lack of comprehension on the part of the one drawing the conclusion. However, we prefer to meet the objection, to go to the very bottom of it, and to show it to be but the hasty deduction of a reluctant mind.

Though St. Augustine enumerates more than once the operations of administration, he does not include among them what, had it a place there, would be one of the most important, the bringing of seminal reasons into actual, visible existence. Let us continue the passage quoted against us. In the objection it is brought to an

[98] Lib., v, 27.

end with the words: "Now He effects all things
through definite intervals of time." Here, nevertheless, St. Augustine's text has a comma only.
It continues closely, as follows, "by which we
see the constellations move from rise to setting,
the heavens change from summer to winter, seeds
in determined space of days sprout, grow, come
to maturity and wither. Animals also within
fixed bounds and courses of times are conceived,
formed and born, and run their career through
the ages of their life to decay and death." [99] The
subject appeals strongly to the Saint's contem-
plative soul; and so, a little further on, we meet
with another and graver enumeration: "He
moves with hidden power His universal creature;
and while angels obey His orders, while the
constellations fulfil their courses, while the winds
rise and fall, while the ocean tosses in billows
that swell and subside under the blast, while
green things sprout and run to seed, while beasts
are born and lead their lives according to their
various appetites, while the wicked are allowed
to trouble the just, the universe revolved by Him
unrolls the ages which He had placed in it, as it
were rolled up, when first it was created. Yet
these, nevertheless, it would not unroll into its
own courses, should He cease to administer by
His provident movement those other things which
He has created." [100] Here the Saint puts before
us in one comprehensive view a summary of
God's providential administration from the first

[99] *Ibid.*
[100] Lib., v, 41.

moment of the visible creature to the last unfolding of the ages. He begins with what is
most elementary, the regular movement of the
heavens by angelic agents, followed by the varied
motions of wind and wave, the movements of
life in the vegetable and animal kingdom, down
to man's actions in the moral order. Yet not a
sign appears of the first passing of seminal reasons into visible existence. And this is the more
remarkable. For there is one thing that he excludes absolutely from administration, the production of any new kind of being. "Should we
think that God now gives being to any creature,
whose kind He had not included in that first
creation, we should contradict the scripture flatly
which says that on the sixth day He completed
all His work. According to the kinds of things
He created in the beginning, He does many new
things, He did not then do. That is clear. But
that He creates a new kind cannot be believed,
since He then completed all things" [101] The passage continued with the enumeration just given.
Naturally, then, had there been any idea of including the visible production of seminal reasons
under administration, He should have begun with
it after terminating the work of the six days with
seminal reasons as such. But not only He did
not do so, but He was also careful not to do so,
for reasons soon to be given. So far, therefore, as St. Augustine is concerned, the Evolutionist finds himself in a dilemma. The assumed
result of Evolution, the production of the first

[101] *Ibid.*

members of a fixed species in the course of time, is either the visible production in actual existence of what was created in seminal reasons in the roots of times, or it is not. If he chooses the former, St. Augustine is against him. Evolution is a process of successive production from seed. St. Augustine asserts again and again that the first individuals of every kind came forth from the earth without seed. If he takes the other alternative, then a new species is produced outside the work of the six days; and this, says St. Augustine, is in flat contradiction to the scripture.

The objection, nevertheless, is urged. According to St. Augustine, the seminal reasons of all creatures to appear at their appointed hours were in the earth in their first creation, as all things of the future tree are in the seed. But the development of these from the seed belongs to administration. Therefore, the production of those also from the earth. Besides, he afterwards makes, as a matter of fact, these seminal reasons the object of administration so far as their visible production is concerned. Here are his words: "Wherefore, creating no further creature, but governing and moving by His administrative act all that He created simultaneously, He works unceasingly." [102]

To the first part of the objection the answer is sufficiently clear. The nature of an analogy requires the analogues to agree in that on which the analogy rests, while they differ in all things

[102] Lib., v. 46.

else. An argument drawn from the points of agreement has its value; drawn from any others, it introduces four terms into the syllogism, and is useless. Here the agreement is in this, that whatever is to be in the tree is in the seed, not from anything proper to the matter of the seed, nor by any material force residing in it, but by virtue of its wonderful power, superior to everything material, of changing water and earth into wood; to account for which we must go back to the creative word that, having created living beings in earth and water potentially in their seminal reasons, brings them into existence without seed. Similarly all things that are to exist, which water and earth produce potentially, were in the earth to be brought into existence, not by any activity in the elements, but without any intermediate action, by the same creative word.[108] From a passage already quoted we omitted a few words not then to the point, but now very much so. Having put before an imaginary questioner Adam in seminal reasons and Adam in his actual existence without concealing the mystery of the matter, St. Augustine continues. "Perhaps he will not understand. For (*in my exposition*) all things he knows have been taken away from his view gradually, down to the materially visible seed. But man did not even reach that when made in that first creation of the six days. There is indeed some likeness as regard seed granted to this matter, on account of those future things bound up in them: nevertheless, before all vis-

[108] *Ibid.*, 44, 45.

ible seeds are those causes" [104] This similitude, then, will hardly serve to found the argument proposed. As regards its second part, the objection assumes that, "The things He created simultaneously He works unceasingly," must have an evolutionary sense. If we remember that for St. Augustine the things created simultaneously, determinations, indeed, of prime matter to future existence, are necessarily non-existent as such, we may ask whether things can be moved and governed before they exist, or whether in speaking of moving and governing, one can ignore the things actually existing and handled and ruled, to consider exclusively the terms of a long series of those operations. Yet this is what St. Augustine, if writing in an evolutionary sense, must have done.

We may begin to gather the solution of the difficulty from the passages lately quoted. Administration is always the specific movement of the creature. It has its term either in the creature itself, or in its reproduction of its own species, not in the production of a term as yet not existing. St. Augustine terminates creation and begins administration with the end of the work of the sixth day. Nevertheless, in discussing: "This is the book of the creature, heaven and earth, when day was made. God made heaven and earth, and every green thing of the field before it was upon the earth, and all the grass of the field before it sprang up. For God had not rained upon the earth, nor was there

[104] Lib., vi, 11.

man to till it. But a fountain was coming out
of the earth, watering all the face of the
earth,"[105] he finds no difficulty in making the
words, "nor was there man to till it" indicate
the end of creation, and, "a fountain was coming
out of the earth," make the beginning of administration.[106] Of the first part of the text he says:
"Not then did He make the herb of the field by
the work in which He now works continually by
means of rain and the cultivating toil of man;
but in that way wherein He created all things
simultaneously."[107] Two things, then, are clear.
When the fountain gushed forth from the
ground, the work of the sixth day as regards
actually existing things was already completed,
and the work of administration had begun. On
the one side creation. On the other administration by rain and human labor, that is, production
from the first individuals watered by the fountain, through the natural process of sowing and
reaping fields watered by the rain.

As to the first individuals watered by the fountain, did their production from the earth belong
to creation, or to administration? The full answer will clear up many obscurities. They certainly came into existence during the course of
time. Earth and sea existed before the creatures
that came from them, and of these there was a
succession, some coming before others. This
St. Augustine not only admits, but indicates very
clearly in distinguishing between the earth, ante-

---

[105] Gen., ii, 4, 5, 6.   [106] Lib., v, 26.
[107] *Ibid.*, 46.

cedent to creatures produced in the day when all things were made simultaneously, not by any interval of time, but by causal relations only, and that same earth prepared to be the habitation of existing things.[108] Because the earth was so prepared, creation did not therefore stop. Its adequate term was not the mere earth and sea, but the world adorned with the first individuals of each kind of life. Those were still in their seminal reasons; and until the last existed, the adequate term of its own creation, the creative act was not finally terminated.

Clearly, then, there was an overlapping of the two orders. The end of creation and the beginning of administration can be considered in two ways. Creation is in the roots of times. Time begins with administration.[109] Hence, absolutely speaking, creation terminated, as St. Augustine insists, with the simultaneous creation of all things, because in it heaven and earth were not created in seminal reasons, but received their actual existence. For them time began, and so, absolutely speaking, administration. Relatively speaking, creation ceases for each species with the existence of its first individuals, and then for each begins administration. The seminal reason becomes the existing being; the root of time passes into the being's own essential time; and so the creature enters into the time of the universe, and takes its appointed place in the order of administration.

[108] Lib., v, 14.
[109] Lib., v, 11.

As determinations of existing matter, seminal
reasons were in the world from the moment of
creation when heaven and earth received actual
existence. If we consider the time of the world,
we say with St. Augustine that from the beginning the world was pregnant of future things
awaiting within it the moment of their bringing
forth.[110]   On the other hand, if we consider
them in themselves without actual existence, they
were not. Still in the roots of their own times,
they were not in time, and so were not subject
to that movement in time and place by which
the unmoved Creator administers the world.
With actual existence came being, time, place,
movement, and so administration. Thus was
completed the creative act, creating all things
together and each in its own time.[111]

Let us now apply what we have seen to the
passage quoted against us: "Wherefore creating no further creature, but governing and moving by His administrative act all that He had
created simultaneously, He works unceasingly."
As creation was perfected formally on the day
when all things were created simultaneously,
heaven and earth, that is, prime matter under its
most elementary forms, in actual existence, vegetable and animal life in their seminal reasons,
no further creation was possible. "Let us not
imagine," says St. Augustine, "from the words,
'the earth was invisible and without form,' any
absence of form from matter, but earth and water

[110] De Trinit., iii, 16.
[111] Lib., vi, 11.

without light, which was not yet created. . . .
The earth is understood to have been called invisible, because it was not yet able to be seen by
reason of the covering waters. It was disorderly,
because as yet it was unseparated from the sea,
undefined by the shore, unadorned with plants
and animals. If such be the case, why were
these visible forms of earth and water, which
certainly are corporeal, created before any day?
Why was it not written: 'God said, let earth
be made, and earth was made; God said, let water
be made, and water was made?' Since it is evident that everything changeable is formed from
some lack of form, this, the Catholic Faith and
solid reason prescribe, that all matter of whatever nature is from God alone . . . whom the
scripture addresses. 'Thou who has formed the
world from unformed matter' [112] The consideration just made persuades us that the words
which, before any enumeration of days, announce: 'In the beginning God made heaven
and earth' express this matter in terms chosen
by a spiritual providence, that heavier and duller
readers may understand it better. With the following words, 'And God said, etc.,' the narration of the order of things formed begins." [113]

Secondly, with the existence of prime matter
under its elementary forms, that is, heaven and
earth, time absolutely began. The material being,
beginning and moving to the determined end of
its course, existed as the measure of the move-

[112] Wisdom, xi, 18.
[113] Lib., i, 27, 28.

ment of creatures yet to be. Hence, the order of administration began, in which God moves and governs unceasingly all He has created simultaneously.

Thirdly, according as other creatures came into existence, they entered the order of administration, and thus the movement and government of creatures in it became more and more perfect.

Fourthly, with prime matter then created under its elementary forms, the first of all the roots of times was quickened into active life, and with it administration, inchoate, it is true, elementary, imperfect, but still administration, was begun. In that matter, but in no way subject to that administration, which directed its simple activity, were the seminal reasons of future things. Having neither existence nor time, they could not be the object of that movement in place and in time, wherein consists the government and administration of Him who is unmoved through time or place. As determinations of passive potency of matter to existences certain to be, they were in the elements. Of that there can be no question. But it is equally beyond question that they were there still subject immediately to the creative act as yet without its adequate term; and until that act should have been terminated adequately by their actual existence, it necessarily excluded all other action. The matter they determined, under whatever substantial form it existed for the moment, shared in the specific activity of that form: as the matter of the thing yet to be, it was utterly independent of that activity. The seminal reasons still belonged to the order of creation. They were still to enter the order of administration.

Fifthly, being in the earth as determinations of the passive potency of matter to existences infallibly future, seminal reasons had their reality. As, therefore, administration of existing beings must take into due consideration future things which are to exist in their own time and place, seminal reasons came under administration in a larger sense, inasmuch as their future existence became a norm directing it. The activity thus exercised in existing things was as regards the future existence of seminal reasons dispositive only, and had not even instrumental efficiency in the actuation of the first individuals of each kind, in which the active element was the creative word exclusively.

Sixthly, with the actuating of the seminal reason, it entered into the order of administration in its own time and place, taking up with regard to other existing creatures the relations decreed by divine providence. Then began for the first individuals of the kind their own individual time, for the kind itself its specific time, and with the unrolling of these processes of time, individuals propagating in each species entered at a fixed moment into the time of the world.

Seventhly, as in this actuation, its root of time passes into the substance's own time, so its seminal reason passes to the substantial form to become the hidden source of all its activity, the invisible seed of its visible seed, linking all its specific vital acts by which it propagates its kind in its own processes of time, to the single creative act of that day in which God created all things simultaneously.

Thus a diligent meditation of St. Augustine's teaching, a striving to learn it as it is, rather than to fit his words to preconceived ideas, opens up a doctrine that, because it is St. Augustine's, could not but be profound, and that in its profundity is worthy of St. Augustine. What seminal reasons were in the order of creation we have learned. We may go on to enquire what they are in the existing creatures, as explained in the treatise *De Trinitate*.

CHAPTER XI

SEMINAL REASONS IN *DE TRINITATE*

WHAT St. Augustine teaches about seminal reasons in *De Genesi ad Litteram* must, we think, be evident to an impartial mind. Those, however, who would have him an Evolutionist appeal to what they hold to be a clearer expression of his thought in *De Trinitate*. Why there should be such a clearer expression in the latter treatise, does not appear to everybody. Generally speaking one would suppose that, were there any difference of clarity between the two, the clearer doctrine should be found in the former, in which seminal reasons are discussed formally and exhaustively, rather than in the latter, which they enter to be touched upon but briefly and incidentally. In the supposition, then, of some contradiction between the two, *De Trinitate* should be interpreted by the teaching of *De Genesi ad Litteram,* rather than the reverse. But the contradiction is no more than a supposition. The two treatises, indeed, view the matter from different standpoints. The matter itself is under different conditions in each. The doctrine involved is perfectly harmonious, as we shall very soon make evident.

Before taking up this task, we must make a remark of the highest importance for the understanding of the question. We are not on the defensive any longer. We have established our position. Even should we fail to conciliate the two treatises, it would not follow that the Evolutionists could claim St. Augustine. Two things are essential to Evolution. First, fixed species are not the immediate term of creation. St. Augustine teaches that fixed species are the immediate term of creation, formal in the seminal reasons, adequate in the existence of the first members. Second, Evolution makes fixed species the result of a long process of successive generations. St. Augustine puts generation absolutely and exclusively into the order of administration, in which, he insists, no new species are produced. This he teaches in *De Genesi ad Litteram,* and there is nothing in *De Trinitate* to contradict it.

In the third book of *De Trinitate* St. Augustine discusses the apparitions of the divinity vouchsafed to man. This, in a way not necessary to explain, brings him to miracles, to the nature of their operation when angelic spirits are instrumental causes, and to the miracles of malignant spirits. Whereupon he says: "Unquestionably in these corporeal elements of the world lie concealed certain hidden seeds of the things that are corporally and visibly born. Of them are some now visible to our eyes from fruits and living things. Others are hidden seeds of those seeds, whence at the Creator's word water produced the first fishes and birds, the earth, the first fruits of their kind, the first animals of their kind. For not then did they so pass into

things thus brought to existence, that the force
in question was used up in the things that were
produced. But for the most part the congruous
occasion of tempered elements, that would enable
them to come forth and accomplish their visible
appearance, is lacking. Behold the smallest slip
is a seed, for, properly planted, it produces a
tree. But of this slip a more subtle seed is a
grain of the same nature, invisible as yet to our
eyes. But now, though we cannot see with our
eyes the seed of this grain, we can infer it with
our reason; for unless some such force was in
these elements, what had not been sown in the
earth would not spring from it so often, nor
would there be born in water and earth, without
union of sexes, so many animals, which, nevertheless, grow and by seminal union produce
others. And certainly bees do not conceive by
sexual union the seed of their young, but, finding
it, scattered, as it were, over the earth, collect it
in their mouths." [114]

Here, then, St. Augustine seems to assert an
active potency of seminal reasons, not only as they
are contained in plants, seeds and other visible generating agencies, but also inasmuch as they are
hidden in the elements of the earth. These seminal reasons are of the same kind as those which
were terminated in the first animals by creation.
They are those which were left over from the
work of creation, and need only a due tempering of things to burst into existence. They produce from the earth what is not sown. They

[114] De Trin., Lib., iii, 13.

are the origin of animals, which, existing without antecedent sexual union, nevertheless, by sexual union reproduce their kind. They are a force, the seed of seeds, even the seed itself of those animals which do not by sexual union conceive the seed of their young.

With regard to the question in which St. Augustine is engaged, he might have explained the miracles of the Egyptian magicians by supposing the evil spirits to have so modified the organ of sight in the bystanders, as to have produced in these the sensation of seeing what had no real objective existence. With this easier explanation he was familiar. Speaking elsewhere of the wonders of the heathen gods, he says: "Most of these deceive the senses by a deception of the imagination, being miracles in appearance only." [115] But as the champion of the literal sense, and the literal sense of Exodus certainly is that the magicians of Pharaoh really changed their rods into serpents and water into blood, and brought forth frogs from the earth, he had recourse to these seminal reasons. "How many men know from what kind of herbs, or of flesh, or from the juices of what plants, put in such a condition, or buried in such a way, or so ground up, or so mixed together, such or such animals will be produced? What wonder, then, if, as the wickedest of men can know whence these or those worms or flies are generated, so evil angels, as more subtle of perception, in the more hidden seeds of the elements, know whence ser-

[115] De Civ. Dei., Lib., x, 16, 2, circa med.

pents and frogs come; and, employing these elements by hidden movements, cause through certain opportune combinations known to them, these animals to be created." [116]   Thus, St. Augustine grants apparently not only an active potency to the seminal reasons in the elements, but also such a force as needs no more than a proper mixing of the elements under definite conditions to produce living beings out of elementary matter.   In other words, he seems to grant all that Evolution demands.   Nevertheless, notice the word "created."   It is the key to his whole mind.

As he is here dealing with seminal reasons indirectly only, while in *De Genesi ad Litteram* he deals with them directly, we might say that his clear doctrine there should be taken as the norm of interpretation for his doctrine here, rather than that this should be set up as a standard according to which that should be corrected. This, however, would satisfy neither the Evolutionists nor ourselves.   We should not be satisfied; for it would seem to imply a real difference of doctrine, where there is none.   Nor would they be content; since their argument is specious, and therefore calls for an adequate answer.   We say, then, first, that its apparent strength comes from their misapprehension of what St. Augustine here terms force.   They take it in the material sense of modern science for the physical and chemical properties in elementary matter, and for what results from them, whatever it be, in vegetative and sensitive life.   St.

[116] De Trin., Lib., iii, 17.

Augustine's idea is quite different. It is permeated with the text of St. Paul, never out of his mind, recurring continually to his page: "Paul planteth, Apollo watereth, but God giveth the increase."

For St. Augustine never wearies of insisting that the real intrinsic reasons of all things are in God alone; that whatever there may be of secondary agency must for its perfect understanding be referred to Him, and that its effects are such only as He grants. Whatever comes into existence is the term of creation, immediate in the order of creation, mediate in that of administration. "To create and to administer creation from the inmost and supreme causes on which all things turn, is one thing; and He alone does this, who is the Creator, God. But it is another to apply some operation from without according to forces and capacities distributed by Him, so that what things have been created may come into existence at this time or that, in this manner or that. All these have been created already originally and primordially in a certain weaving into one of the elements, but they come into existence as opportunities are afforded. For as mothers are pregnant with their offspring, so the world itself is pregnant with the causes of things to be, created by that supreme Essence wherein nothing springs to life or dies, nothing begins or ceases to be. To apply exteriorly causes that present themselves, which, though not natural, are applied acording to nature, so that what hidden things are contained in nature's secret recesses may break forth and be created exteriorly by unfolding in a way their measures, numbers, and weights which in secret they received from Him who disposed all things in measure, number and weight, is within the power, not only of evil angels, but also of wicked men, as I showed by the example of agriculture just now." [117] This example of agriculture is the continuation of the passage quoted in No. 114; "For the Creator of invisible seeds is the Creator of all things; for the things coming into existence before our eyes, receive from hidden seeds the first beginnings of their coming forth, and receive, as from original rules their growth of definite magnitude and their distinctions of forms. As therefore we do not term a man's parents, his creators, neither are husbandmen creators of the crops, although by their movements applied externally the power of God works interiorly what things are to be created." [118]

Let us remember that St. Augustine is discussing false miracles in the order of administration. His doctrine is perfectly clear. In the order of administration no new species comes into existence. God operates in the perpetuation by generation of those originating without seed by creation. In this He uses intermediate agents that work in their own specific times. As such, they receive from Him the power of working in their own seminal reasons now become active

[117] De Trin., iii, 16.
[118] *Ibid.*, 13.

because existing; for all future existences were determined in the first creation individually to their specific nature in their seminal reasons. But for their action they need matter passively determined in that first creation to become specifically what they themselves are. Their operation is extrinsic, dispositive only. But they dispose the matter naturally, by the application of their specific movements. The principal cause of the effect is God creating all things in seminal reasons. Recall St. Augustine's frequent example of the vine. It springs from a cutting or from a seed. It grows by changing earth and water into wood. It fructifies by continuing this change into the grape, at first sour, then sweet and full of juice. This, when pressed out, changes by the power received from the vine into wine. But all this is material. Neither earth, nor water, nor seed, nor cutting, has in itself the power to perform this miracle of nature. For the origin, nay, for the exercise of these powers in the vine, for the power of responding to them, for the actual response in the elements, we must go to the first creation of the seminal reason in matter, to the administration by God of that seminal reason, existing united with matter. "My Father worketh hitherto, and I work." What He works through processes of time by the instrumentality of the vine, He can do instantaneously by His own power, and work the miracle of changing water into wine.

Between the miracle which God works alone, exercising His supreme dominion over all mat-

ter, changing it from form to form instantly by
the mere act of His will, and the natural process,
whereby in a definite period of time He accomplishes the same effect through agents working
according to their nature in the natural order,
comes in another class of miracles. In these, as
in the natural process, no less than in the direct
miracle, God is always the principal agent, the
sole intrinsic efficient cause. But to perform instantaneously those external movements which,
spread over a long time, are the generating agent's
part in the production of new being, He may
use the extrinsic ministry of angels, whose subtle
intelligence penetrates swiftly and deeply into
the secrets of nature, and whose powers are in
their exercise independent of material means.
Thus, working according to their own nature, and
according to the nature of things, though not in
the natural order, they will make the tempering
of elements in number, measure and weight,
which is the natural function of the generating
agent, so disposing the matter to specific life in
this or that individual, which, as an actual result, is the work of the Creator, creating in matter seminal reasons in the beginning, and in time,
that is, in the order of administration, actuating
them, through the instrumentality of generation
or of its equivalent. What God can command
the good angels, He may in His providence permit to the evil angels, who thus collected, tempered, disposed the matter destined from all
eternity to be the material term of His divine
operation in this producing of frogs under the

magician's hand. Thus, we see the force of the expression, "caused these animals to be *created*," to which I called attention a moment ago.

Two ideas are always present in the theory of Evolution. One is of its very essence the production or differentiation of species by successive generations. The other belongs rather to its apologetic, namely, the economizing of divine activity. Both are utterly foreign to St. Augustine. Not only can they not be derived from anything in this treatise, *De Trinitate*, but they cannot even be made to fit into it. The Evolutionist restricts divine activity in the administration of material creation to conservation and co-operation. Were one to tell him that he makes the former purely negative, namely, abstention from destroying what has been created, and distinguishes the latter from the former but in name, he would be indignant. Nevertheless, such is the case. To make conservation consist merely in the abstention from destroying His creature, and co-operation, the leaving of the creature to the work of its own proper faculties and force, flow logically from the notion that the contradictory of Evolution is found in successive creations, the getting rid of which should be for Christians the Evolutionist's chief praise, as for the rest of the world it is the getting rid of the Creator altogether. When the Evolutionist speaks of successive creations, he considers them in themselves, not in the terms of creative act. This is undeniable; otherwise his argument would be pointless. Moreover the variant he uses: "Divine intervention must not be admitted without necessity," declares it. But such intervention, which has a perfectly legitimate sense when there is question of the extraordinary miracle, must, in the ordinary course of things, necessarily suppose a divine activity suspended, interrupted, beginning, ceasing, beginning again. It assumes that the Creator is normally inactive, and that His creatures must get along as best they can without Him, and that there is no clearer sign of a devout, reverential spirit than to ignore Him in His works. That conservation and co-operation are in the Creator undistinguishable from the one, eternal creative act, and in the creature are simply its necessary, unbroken extension: that there is no movement in the creature but what is begun, continued, completed by Him, who, as we have heard St. Augustine saying, unmoved in Himself, moves all things in time and place, are ideas incompatible with the Evolutionist's "successive creations." If this means anything, they are sound without sense. If they express the mystery of St. Paul's words: "In Him we live and move and are," [119] "successive creation," is a figment of the imagination. Yet, as they lie at the foundation of the treatise *De Genesi ad Litteram*, so are they the key to this explanation of miracles in this treatise *De Trinitate*.

According to St. Augustine, God is no less the agent in the ordinary processes of nature, than in the extraordinary miracle. In each His activity is essentially the same, and differs in mode

[119] Acts, xvii, 28.

only. "By the divine power administering the whole spiritual and corporeal creature, the waters are called from the seas on determined days in every year to be poured over the earth: when, without previous signs of gathering rain, such floods followed so swiftly the prayer of Elias, the divine power was manifested to those to whom the miracle was given. So God works the lightnings and thunders that ordinarily occur: but, because on Sina they were made in an unusual way, they appeared there most evidently as signs. Who draws water to the grape-cluster through the roots of the vine, and so makes wine, but God, who, though men plant and water, gives the increase? But when at the Lord's command water was in a moment changed to wine, even fools acknowledged that the power of the divinity was shown. Who but God clothes the tree year by year with leaf and flowers? But when the rod of Aaron the priest blossomed, the Godhead spoke in a way with doubting mankind. Surely the earth is the common matter of the generation and conformation of all trees and of all animals; and who makes them but Him, who commanded the earth to bring them forth, and in His same word rules and moves what He created? But when He changed instantly and quickly the same matter from the rod of Moses into the serpent's flesh, there was a miracle, of a thing changeable, to be sure, but, for all that, an extraordinary change. Yet who animates every living thing that comes to birth, if not He who, as need had arisen, gave momentary life to

that serpent? And who restored the souls to the
corpses, when Ezechiel saw the dead arise, unless He who gives life to flesh in their mothers'
wombs, that they may be born to die? But as
these things occur in, as it were, the unbroken
stream of things slipping onward in their flow,
and passing in their familiar course from their
concealment into the light, and from the light
into hiding, they are said to be natural: when,
however, by an unwonted mutability they are
forced upon men for an admonishing, they are
given the name of mighty works." [120] The nature
of a miracle, therefore, is to be extraordinary and
rare. As regards miracles, then, St. Augustine
would admit the term, "intervening," since, of rare
occurrence, they are only to be used to explain
facts incapable of natural explanation. But in
that other sense which divides the creature's operation from the Creator's, requiring the former
to be left to its own natural powers and admit-
ting its Author only when it is at a deadlock,
the principle would have been for him not so
much meaningless as blasphemous.

It is unnecessary to prove what we have already pointed out, that the differential generation
of new species in the order of administration,
though essential to Evolution, and utterly rejected in *De Genesi ad Litteram,* finds no support in this treatise. There is not in it the faint-
est suggestion that the wine, the rain, the blos-
soming rod, the serpents, the frogs, were not of
the same species as those coming into existence

[120] De Trin., Lib., iii, 11, 12.

in the ordinary course of nature, or that the flies, bees, plants, supposed to arise without seed, have not been constant in their species during the whole order of administration; while there is much suggested and even said in the contrary sense. We shall close this chapter, therefore, with a summing up of St. Augustine's teaching concerning seminal reasons in this treatise *De Trinitate*.

Seminal reasons, the seeds of things, lie concealed in the corporeal elements of the world. They come to exist actually by (1) creation, (2) ordinary generation, (3) directly from earth and water, (4) by miracle effected, (a) directly by God, (b) indirectly by ministry of angels. With regard to creation there is nothing more to be said. The matter has been fully discussed. The miracle effected directly by God differs from creation in this only, that it is in the order of administration, since what it effects exists already in species; and from ordinary natural generation and its analogues, in that it accomplishes in a moment what they do in regular processes of time, accomplishing it often in subjects in which they could do nothing.

In ordinary generation the seminal reason is in the agent the invisible seed of the visible seed, the virtue created by the creative act, actuated in the existence of the first members of the species, to be transmitted by them in the generating of others like to themselves. Here it has become active, because the potency it determined has been actuated by the form. In the matter on which

the agent acts it is always that determination of
the passive potency of matter, which is the formal
effect of creation, whereby God is formally Creator of all things, whether their existence be the
adequate effect of the word of God only, or the
result of natural operations or of miracles. The
active seminal reason in the agent is the efficient
cause, the passive seminal reason in the matter
is the material cause. Both depend ultimately
on the creative word. The efficiency of the visible generating agent or agents is the conjunction of active and passive seminal reasons under
conditions necessary for the production of the
effect. The material seed is but the subject of
the active seminal reason.

As regards miracles performed by the ministry of angels. In the first place these were
foreseen and decreed in the universal order of
providence. Their matter was determined in the
first creation in seminal reasons, to this extraordinary mode of actuation in an order outside of
and superior to the order of nature. This is essential, the necessary consequence of St. Augustine's insistence on creation in definite numbers
and times. The active principle of production is
the divine Word, the ministry of angels is instrumental only, commanded if they are good, permitted if they are bad. They collect the matter
determined by seminal reasons to this particular
effect at this determined time. They mix it in
suitable proportions, they provide the suitable
temperature, as do the generating agents in ordinary generation. But God works the effect in

the extraordinary way. The active potency of the seminal reason is absent: the passive potency only of determined matter is there.

What St. Augustine held with regard to the supposed production of plants and insects from earth and water is sufficiently clear. Whether there be question of creation, generation, miracles, or this production from the earth, the passive seminal reasons are always the same, the agency of creatures is always instrumental, the principal efficient cause is eventually always God. To understand him we must realize, as he did, the intimate nature of God's operation in creatures, a notion so alien to the modern mind. Not the smallest particle of matter, not a creature, however minute, escaped His providence. Not only was the potency of definite matter determined in the moment of creation to definite beings, but the defining of the choice of matter was ruled by that providence which would bring the matter so determined to be at its proper time in such a place that the instrumental causes disposing it for actuation might act on it. With this in view, he says that not all the seminal reasons were exhausted in the creation of the first individuals of each species. Many through lack of congruous tempered elements remain inactuated. From others are produced those creatures coming from water and earth, their matter being providentially disposed with its seminal reason so that under the ordinary course of things would occur its proper mixing at the right temperature, the general instrumental efficiency of the generating agent. But this generating agent lacking, and with it the active seminal reason, the seed of its seed, the divine Word is the immediate cause. Nevertheless, for reasons already given, this is not creation. The species is always fixed, therefore already existing: the process belongs to administration.

That this is his doctrine is clear. As regards the coming into existence of seminal reasons he makes no distinction of material conditions, the activity of elements congruously disposed and mingled. "All these things were indeed created originally and primordially in a certain weaving of the elements into one; but they come into existence as opportunities are afforded." [121] It will not do to say that he is speaking here especially of the order of administration. This might explain the turn of the phrase; but the assertion contained is universal. We must, then, see what this "weaving of the elements into one" is. Nor need we go far to discover it. It is the application of some operation from without, according to forces and capacities distributed by God.[122] What the weaving together is physically we shall see hereafter. It is formally the execution by divine providence of the actual order of creation, chosen by God's will to exist, and decreed down to its last particular. In this decree was contained, not only what was to exist, but also the when and the how.[123] And thus was all creation, and in it all the elements, woven into

[121] De Trin., iii, 16.
[122] *Ibid.*   [123] *Ibid.*

one. Viewing the matter physically, this application of the elements is extrinsic, not to the adequate effect, as is obvious, but to the operation of creation, direct or indirect. It gives in no way that determination of the form which Evolution demands; for this is the effect of the operation of creation in its intrinsic nature. On the contrary, all that St. Augustine grants it is the determination of the time and the manner of the creature's coming into existence. The question now arises —Does this application of the elements determine time and mode principally or instrumentally? St. Augustine calls the "weaving" in the elements themselves, a "tempering." "Seminal reasons do not come into existence, because of the lack of suitable occasions of things tempered." [124] He gives as an example of determination of time, the due tempering of heat and cold which makes summer more propitious for certain generations than winter: [125] and as an example of mode, Jacob's method of determining the color of the future lambs of Laban's flock.[126] Hence their action is instrumental to the fulfilling of a condition. It gives the occasion but in no way enters into the production of things.

Nevertheless, this tempering of the elements, condition though it be, and primordially to be referred to the Creator, differs necessarily in its proximate causes according to the different modes of production. In generation it comes chiefly from the generating agents themselves. In the

[124] *Ibid.*   [125] *Ibid.*, 17.
[126] De Trin., iii, 15.

miracle in which angelic spirits have no part either commanded or permitted, it is part and parcel of the perfect obedience of material things to their Creator. When angelic agencies come in, it may be the result of the ordinary course of nature in the elements themselves perceived and taken advantage of by the subtler faculties of spirits, or else it may be brought about by combinations and mixtures deliberately effected by these.[127] While in production from earth and water, it would be the result of the ordinary course of nature under divine providence.

We must observe that to this activity of the elements St. Augustine grants a disposing function, small and subordinate, it is true, but for all that, real, even in direct creation. One may ask, how this is to be reconciled with the pure passivity of matter we have constantly insisted upon? The answer is not difficult. In creation pure and simple St. Augustine asserts the pure passivity of *seminal reasons,* as a mere determination of prime matter to being that is actually to exist. All the activity entering into creation is that of the creative act, under which the seminal reasons lie immediately until it shall have its adequate term in their actual existence. He does not assert the pure passivity of *the elements,* which would be a contradiction. Existing elements must have their substantial forms, and consequently their active potency. But though the matter of the seminal reason must exist under some form, he ignores the activity of any such form, as entering neither

[127] *Ibid.,* 17.

formally nor adequately into the concept of creation. This is clear. Recall the conclusion we reached in Chapter x. "Seminal reasons were in the matter, but subject immediately to the creative act, which until terminated adequately, must exclude all other activity. The matter determined might have its own activity from the substantial form under which it was existing: as the matter of the thing to exist it had none. Seminal reasons still belonged to the order of creation; they were still to enter the order of administration." [128] On the other hand, as matter existing in the order of administration was to lose its form, to receive in the order of creation that to which it was determined, it might well do that under the activity of other elements.

Hence, for the first production of the beginnings of every species we must admit a tempering of the elements. How far this was the effect of natural forces working according to divine administration; how far the work of ministering angels; or whether it is to be attributed exclusively to the obediential potency of matter, are here useless questions, since such tempering must have been dispositive only and could not enter into the formal production of things. On the other hand, it suffices amply to explain literally the words: "let the earth bring forth the green herb; the waters, the creeping creature having life; the earth, the living creature in its kind." For whatever be the immediate tempering cause, its immediate effect in nature is to deprive of its existing form that mat-

[128] *Supra*, pp. 98, 99.

ter which at this moment can receive no other
form in natural course, since it is ordained now
to the actualization of its determined potency as
a seminal reason. Hence, it has for the moment
no place in nature, being thrust out of the administrative order into the creative, from natural
transformation to that unique information, which
is the beginning of nature. Truly the waters and
the earth *"ejiciunt"* the creature. Yet, St. Augustine never forgets that the passage from the
seminal reason to the existing being, the formal
production of the material being, first of its kind,
is not the effect of the activity of elementary
forms, but of the creative word, to which is due
both the seminal reason and also its adequate perfection. [129]

But St. Augustine says distinctly that the force
is in the elements, "Unless some such force was
in the elements that would not generally spring
from the earth which had not been sown there,
nor would so many animals come into being in the
earth or in the water without union of sexes.
Nor would bees exist." [130]   The objection always
returns to the same point. St. Augustine is not
speaking of physical force, but of seminal reasons, the formal term of creative power, and as
such to be termed force, as regards things to be.
He speaks of it as being in the elements, because,
engaged with the order of administration, he sees
that prime matter with all the determination of
its potency to things yet future, exists, nevertheless, at the present moment under some elemen-

[129] De Trin., iii, 13.   [130] *Ibid.*

tary form. But though in the elements, seminal reasons viewed as force, are not of the elements, so that by any efficiency of the formal active potency of earth or water, the production is effected of the things produced without seed. They are there as seed is in its receptacle. Some disposition of the receptacle is needed for the seed to germinate, but the germination is due to seminal reasons, the virtue of which is in the seed. The creatures produced from earth and water are, as we learn from another place, flies.[181] Of them some were supposed to be generated in decaying vegetable matter. But not to the elementary forces there at work would St. Augustine grant any efficiency. He will find the seminal reasons of those flies in the earth itself. Thither he carries back their origin through the decaying plant matter, by means of his doctrine of vegetable life, intermediate between inanimate matter and sensitive life. "The vegetable kingdom, joined on to the earth in a continuity by means of its roots, and brought forth, not only before the animal world, but also before the heavenly bodies, as soon as the dry land appeared, is so intimately connected with the earth and water, that the flies in question may, without absurdity, be included in the number of those minutest bodies coming directly from the water and earth." [182] Those minutest bodies, as we have seen, originate, as the term of the creative word, from seminal reasons unactuated in the adequate creation.[188]

[181] De Civit, Dei., lib., xvc, 27, 4.
[182] De Gen. ad Litt., iii, 23.　　[188] De Trin., iii, 13.

In this part of *De Trinitate* St. Augustine is occupied with the false miracles by the Egyptian magicians. His object is to show that in the working of them, these magicians exercised no power over nature. Whatever they did by means of their familiar spirits was instrumental only and dispositive: the production of serpents and of frogs was the work of the Creator, permitting the magicians' evil art. This production he assimilates to the production without seed in this present order of administration of the living beings we have just been discussing. This, assumed to be the natural way of producing flies, has for its cause the seminal reasons in matter acted upon directly by the divine Word instead of indirectly through seed. For its condition it needs a certain tempering of the elements. For the miracles in question the seminal reasons are in the earth. The ministry of spirits effects the tempering of the elements. The actual effect is the work of God. Thus, there is a kind of analogy between the supposed production of flies from the earth without seed in its relation to ordinary generation, and the miracle by the ministry of angels in its relation to the miracle purely divine.

In the explanation, nevertheless, of such generations assumed to be without seed St. Augustine felt, without doubt, that he was facing grave difficulties. Could he have divined that the opinion of the day erred as to the fact, and that there is in the order of administration no natural exception to the general law that there is no coming into existence without seed, he would have been greatly relieved. However, other ideas were in possession; and, conforming to them, he worked out legitimately his doctrine of seminal reasons. That his tone is sometimes doubtful, is not surprising. Indeed, it is worthy of remark that in this matter he does not in things difficult of explanation appeal, as is his custom in others certain by revelation, to the eternal truth of God. Why he did not we will not attempt to decide. We do not claim for him a keener insight into things than evidence warrants. It is clear, nevertheless, that, whatever might be the fact, generation without seed was for him a particular case confined to creatures the most insignificant exciting no idea of any evolutionary development. In it flies, worms, bees, always recur specifically the same as their predecessors, just as though they had been generated by seed. On this to found the theory that St. Augustine was an Evolutionist, or that his doctrine favors Evolution, whether of Darwin or of any other, in the least degree, is to transgress every law not only of interpretation, but even of reasoning.

What seminal reasons mean in the treatise *De Trinitate* is clear. In it there is question of administration only, in which the activity of secondary agents replaces the immediate activity of the creative word. Yet in themselves these agents with their material forces were powerless to determine the potency of matter. This virtue as far as they received it, they received from the creative word that, in the order of creation, perfecting by actual existence the potency of matter which it had determined, put into material things the power of generation, of production, which St. Augustine calls the hidden seed of seed. This, then, taken in conjunction with the passive potency of definite matter ordained to its operation, both being the exclusive effect of the creative act, constitutes the seminal reason, active and passive in the order of administration, by which the world is pregnant of all that is to be, verifying absolutely the texts ever on St. Augustine's lips: "Paul plants; Apollo waters, but God gives the increase." "My Father worketh hitherto and I work." So also does St. Thomas, by right of intellectual succession, the legitimate interpreter of the Holy Doctor, understand the matter.

CHAPTER XII

ST. THOMAS AND SEMINAL REASONS

TO reach an understanding of the Angelic Doctor's mind with regard to seminal reasons, we must distinguish in St. Augustine his general doctrine in *De Genesi ad Litteram*, as a literal exposition of the history of creation, as a conciliation of the six days with the one day in which God created all things simultaneously from this particular point in it. To the former St. Thomas does not commit himself. In that matter he is not a commentator; and even if he were, there are other opinions entitled to respect. He remarks, nevertheless, that though St. Augustine's teaching regarded as an explanation of the text of Genesis differs greatly from that of other saints, as regards the production of things there is no great difference between the former teaching the simultaneous potential creation of all things, and the latter holding to their successive production. Both agree in this, that in the first production of things prime matter was under elementary substantial forms, and that animals and plants were not actually existing. They differ on four points. First, as to whether there was a space of time in which there was no light; in which, secondly, the formed firmament did not exist; in which, thirdly, the earth remained covered with water; and, lastly, in which the luminaries of heaven were not formed.[184]

This assertion of St. Thomas must come as a surprise to one possessed with the idea that seminal reasons are a special concept of St. Augustine, and as a surprise still greater to those who maintain that St. Augustine's doctrine resting upon them must find its logical conclusion in Evolution. The fact is, as we shall show, that there was not in the mind of St. Thomas the slightest question about the seminal reasons in the full sense of St. Augustine. On the contrary, to any clear idea of God's operation, whether as Creator or Administrator, in this world, with its determined species and its definite number of individuals in each, of the origin of each in its first members, of the mysteries of generation, they are absolutely necessary. Whatever may be said about the term, the reality was common, not otherwise than as understood by St. Thomas himself.

"Avicenna," says St. Thomas, "held that all animals can be generated in nature's way without seed by some mixture of the elements. But this seems out of accord with nature, which proceeds to its effects by determined means. Hence, what is naturally generated from seed, cannot be naturally generated without seed." [185] Having laid down this principle, he proceeds to state his opinion in explaining the words: "Let the waters produce the creeping creature having life, and the fowl flying over the earth." "In the natural generation of animals the active principle is for-

[184] i, lxxiv, 2.0.   [185] i.lxxxi, ad 1am.

mative virtue in the seed for things generated from seed. In place of this for things generated from putrefaction is the virtue of the heavenly bodies. The material principle in the generation of both is some element, or something elemental. In the first beginning of things the active principle was the word of God, which from elementary matter produced animals actually, according to other saints; or virtually, according to St. Augustine. Not that water or earth had in itself the power of producing all animals, as Avicenna supposed, but because this very fact that by virtue of seed or of the stars animals can be produced from elementary matter, is from the virtue given to the elements from the beginning." [136]

According to the Evolutionist, St. Thomas makes St. Augustine teach that for the production of fixed species the material part is generation by seed, or by the virtue of the heavenly bodies, the formal differentiating agent, distinguishing each new being in its species from all that went before, is a special power given to the elements from the beginning. That this virtue or power here mentioned is the seminal reason is undisputed. Therefore, St. Thomas confirms the statement that St. Augustine was an Evolutionist.

But St. Thomas says that the difference between St. Augustine and the other Fathers is reducible to four small points, in none of which are seminal reasons ever hinted at. If, then, his doctrine in this matter makes St. Augustine an Evo-

[136] *Ibid.*

lutionist, all the Fathers are, according to St. Thomas, the same. All must agree that the seminal reason is the active principle of differentiating species, whereby from a few primordial ancestral types we reach the vast variety that in animal and vegetable kingdom both by land and sea enriches the existing world. Yet strange to say, no clear sign of any such doctrine is to be found in their writings, where there is much to the contrary. To get the semblance of a sign means violence to their text.

So foreign is such a notion to any accepted idea of the teaching either of the Fathers or of the schools, that we must offer another interpretation. We may save time and space by leaving out the special case of generation without seed, and say that St. Thomas recognized but two active productive principles, seminal virtue in natural production, and the word of God in the creation of the first members of each species. In both productions the passive principle is prime matter under some elementary form actually or equivalently. But how can seminal virtue acting upon matter produce a being specifically identical with the parent plant or animal? This question, commonly overlooked, presented itself forcibly to St. Augustine and St. Thomas, who saw that unless there were a corresponding receptivity in matter, seminal virtue would be without effect. A determination in this definite matter to this particular substantial form is necessary. This determination was given according to St. Augustine in the first creation of all things in potency: according to the other

saints, when the first of each species was created
in act to reproduce its kind. In a word, it was
the seminal reason. That in receiving this determination the passive potency of matter remains
still purely passive, St. Thomas indicates by his
deliberate change to the passive voice. He does
not say, "the very fact that elementary matter
can produce animals by virtue of seed is from a
virtue given to the elements in the beginning";
but, "that animals can be produced from matter."
Why, though passive, it is called a virtue, we have
explained fully.[187]

Elsewhere the question of seminal reasons is
raised; and in explaining the propriety of intro-
ducing them into the discussion of matter's function in creation, St. Thomas notes two differences
between God's operation and that of an artificer.
The first regards the matter of the work. The
artificer takes material disposed to receive the
form he will give it. God does not take, but creates; and in creating matter creates in it the disposition to receive all the forms that are to exist.
That is, He places in matter passive seminal reasons of all future beings to be produced from it.
The second difference touches the form. The
artificer can not create a form capable of reproducing itself specifically. God creates such
forms, that, informing matter determined to their
reception, actuate it with active potency to reproduce the species. Thus passive seminal reasons
in their subject become active. This potency put
by God in matter to receive whatever He dis-

[187] *Supra*, p. 57.

poses is called, *obediential;* and as such extends
to the supernatural and the natural order, including miracles as well as natural effects. According as natural effects follow, it is, with regard to
them, their seminal reasons.[188]

As St. Augustine, so St. Thomas distinguishes
absolutely creation from administration. By the
former nature was so constituted that the beginnings of life then created should subsist in themselves, and from them others should be propa-
gated. Thus, it received active and passive virtues,
which Augustine calls seminal reasons. These are
two-fold. Some are common, moving to every
species, as heat, cold and the like. Others
move to determined species, as the seed of a
lion or of a horse. The common active and passive virtues are given by the work of the first
three days, called by the Holy Doctors, the work
of distinction, that is, the creation of light, the
division of the waters above the firmament from
those below, and the gathering of these into one
place so that the land appeared. Considered as
a constant separation, these belong to administration. As creation, they consisted in the creation
simultaneously with the elements of the common
active and passive virtues, that we today call
physical forces. In the consequent days is found
the work of ornament, the filling the earth with
various kinds of life. This again, considered as a
constant succession of such, is a work of administration. In that of creation it was the confer-
ring on the first species their specific active and

[188] 2 Dist., xviii, 1, 2.0.

passive virtues. It is evident that the common active and passive virtues formed in elementary matter as such, having their formal effect, distinction, moving to every species in general, must be insufficient for the production of definite life, which finds its beginnings in specific seminal virtues.[139]

But one may ask, what are these seminal reasons in their reality? St. Thomas answers the question in the article we were considering lately. "Some say that the form of the species is not received in matter otherwise than by means of the form of the genus; so that fire is fire by another form than that by which fire is a body. Therefore, that general incomplete form is called the seminal reason, because on account of such form there is in matter a certain inclination to receive specific forms. This, however, does not seem true; because every form that follows substantial being, is an accidental form. . . . Nor does this agree with what Augustine means; because the special form does not follow necessarily from the virtue of the general form. Wherefore such a virtue is not one according to which something is necessarily made, but according to which it may be made.

"Therefore others say that, since all forms, according to the Philosopher, are educed from the potency of matter, the forms themselves must pre-exist in matter incompletely according to a certain beginning, as it were; and because they are not perfect in their being they have not a perfect

[139] 2 Sent. Dist., xiii, 1, 1.

power of acting, but one incomplete. Consequently, they cannot go forth into act by themselves, unless there be an external agent to start the incomplete form to act, so that it thus co-operates with the external agent. Otherwise generation coming solely from without would not be a natural change, but violent. They call those incomplete virtues, pre-existing in matter, seminal reasons, because they are according to their being complete in matter as the formative virtue is complete in the seed. But this does not appear true; because, though forms are educed from the potency of matter, that potency of matter is, for all that, not *active* but *passive* only. . . . For as in simple bodies we do not say that they are moved of themselves as regards place, since fire cannot be divided into *moving* and *moved,* so also such a body cannot be altered of itself, as if some potency existing in matter should act somehow on the very matter in which it is, by educing it into act. But both these happen in living beings. They are moved locally and altered of themselves on account of the distinction of their organs and parts whereby one is moving and altering, another, moved and altered. Therefore, seminal virtue is not to be understood in other things as in those possessing life. Neither does it follow, if potency in matter is only passive, that generation is not natural. Matter does its share in helping to generation, not by acting, but inasmuch as it is apt to receive such action; and this aptitude is called the appetite of matter and the beginning of the form. . . . I grant, therefore, that in matter there is no active potency but purely passive, and that we call active virtues complete in nature with their own passive virtues, as heat and cold, and the form of fire, and the virtue of the sun and such like, seminal reasons, and that we call them seminal, not because they contain the being imperfectly, as in the case with the formative virtue in the seed, but because such virtues were, by the work of the six days, placed in the first created individuals of things, so that from them, as from seeds, natural things might be produced." [140]

In this idea of seminal reasons St. Thomas excludes from matter any active potentiality, whether that of generic or of inchoate forms. This at once excludes whatever activity Evolutionists demand to be terminated in the permanent species. For since all activity comes from the form, and is determined in its specific virtue by the form, such activity would necessarily be reduced to the inchoate permanent form, the containing of the being imperfectly in the manner of seed. This will appear clearly to one considering any evolutionary theory proposed as in harmony with Catholic teaching. Secondly, seminal reasons are the active virtues complete with their passive virtues. "According to Augustine we call seminal reasons, all the active and passive virtues placed by God in creatures, by means of which He brings natural effects into being. . . . Wherefore among these seminal reasons are contained the active virtues of the heavenly bodies, which are nobler than the active virtues of lower bodies, and so are

[140] 2 Sent. Dist., xviii, 1, 2.0.

able to move them. They are called seminal
reasons, inasmuch as all effects are, as regards
origin, in active causes, as in, so to speak, seeds." [141]
Thirdly, the passive potentiality of matter, in
the abstract universal and negative, remotely
capable of every actuation, with no proximate relation to any, when determined to some definite
future form, though it remains passive, becomes
particular and positive. Though "purely passive," it acquires "an aptitude" to such forms, to
the exclusion of others, that is called its "appetite" for them and the beginning of the form.
This is most important; since as we have seen,
and shall see again, it is the key to St. Augustine's creation in seminal reasons and is so understood by St. Thomas, enabling him to say against
future Evolutionists and all others who would put
that Saint in substantial opposition to the other
Fathers, that the difference between his doctrine
and theirs, regarding the production of creatures,
is in a few matters of minor moment only.

One may ask, why St. Thomas insists so much
on the active virtues in seminal reasons, while St.
Augustine has so little to say about them in *De
Genesi ad Litteram*. The answer is obvious. He
is not interpreting Scripture. The problem of
how to conciliate the one day in which all things
were made simultaneously, and the succession of
the operations of the six days, is therefore outside
his scope. His concern is to consider creatures
actually existing, the term of the creative act,
and beginning their natural operations in the order

[141] De Verit., v, 9 ad 8m.

of administration. This brings in necessarily into his discussion the active virtues, as it brought them into St. Augustine's *De Trinitate*. Another question arises regarding the expression: "The active virtues complete with their passive virtues." This may be expressed more fully thus: Active virtues with that complementary receptivity of the material element assigned to each by the Creator, so that they may produce the concrete effects called for by His providence, as it rules this order of creatures. It is clear, and St. Augustine's teaching that St. Thomas makes his own rests upon this truth, that mere activity can no more produce an effect without a passive potency apt to receive its formal action, than that such a passive potency, however apt, however determined, whatever its "appetite for the form," can pass into being without a definite concrete virtue to actuate it. Ignore this, and you are engaged in mere abstract speculation. But the seminal reason is supremely concrete. It contains the thing that by consequent necessity must exist. It is so identified with it, that creation in the seminal reason is creation of the thing itself, as St. Augustine saw so clearly. "When all natural causes concur so as to make one perfect cause, the effect follows necessarily unless something occurs to hinder. And this is the sense of Augustine." [142] But such is the relation of seminal reasons to the providential working out of the order of creation, that such hindrance is impossible.

Here we meet another difficulty occurring under

[142] 2 Sent. Dist., xviii, 1, 2, ad 5ᵐ.

various forms to those who dissent from our explanation of St. Thomas and St. Augustine. It may be expressed thus: If in matter there be passive potency only, to say that seminal reasons are in matter is a contradiction; since what is seminal is naturally active. St. Thomas answers that, as in actual generation the passive principle is necessarily included, its ideas may enter into the notion of seed taken in its fullest sense.[148] This returns to what we have just pointed out, that the seminal reason is something definite, concrete, identified with the thing that is to be. Though from the nature of things it is expressed in terms of the potencies, it is not the active potency or the passive considered in the abstract merely as such, but taken together, as they certainly shall be, to produce the thing decreed by God to exist. The former view would never take us beyond possibility; while the question is of actuality none the less real because future, since its existence is absolutely certain. So St. Thomas and St. Augustine consider them in natural production. In creation the word of God replaces the natural active principle; and as this word is above all nature, and creation is neither natural nor supernatural in strictness, but rather the beginning of nature for the creature that is its term, we find in nature when creation is in question, only the determined passive potency as the seminal reason of that creature. Hence, St. Thomas says: "A thing is said to preexist in creatures according to causal reasons in two ways. One is according to

[148] i, cxv, 2, ad 3m.

active and passive potency, so that not only can
it be made out of preexisting matter, but also that
some preexisting creature can do this. The other
is according to passive potency only, namely, that
from preexisting material it can be made by God;
and in this way, according to Augustine, the body
of man preexists in created works according to
causal reasons." [144] For this reason we insisted
that, for an adequate idea of St. Augustine's doctrine of creation in seminal reasons, we must
never separate in our thought the passive potency
of matter determined by the Creator to things
that are to exist, from the creative word so determining them, under which it lies until in obedience to that creative word, creation finds its adequate term in actual existence.

To return for a moment to a passage lately
quoted, in which St. Thomas expounds in his own
way the doctrine of seminal reasons [145], will con-
duce greatly to a clearer perception of his perfect agreement with St. Augustine. Though he
denies that in heat and cold, in the form of fire,
in the virtue of the sun the future being is to be
found even in that imperfect manner which he
grants to the formative virtue of seeds, he calls
them, nevertheless, seminal virtues. This may, at
first sight, seem inconvenient. We have learned
to consider the formative virtue of the seed as
seminal virtue just because it contains in itself,
however imperfect the manner, the future being.
One may, therefore, ask, how we give those

[144] i, xci, 2, ad 4m.
[145] *Supra.*, pp. 134, 135.

agents that in no way contain the future being, the same title. He may then object: "Either there is a contradiction in St. Thomas, which here cannot be granted; or else, as is more likely, you have not grasped his doctrine, which must recognize in those forces a real causality in the production of the first specific life, not a mere dispositive agency." But we must note that all predications in creatures of seminal reasons are analogical only to the adequate sense, which, transcending the finite order of creation, may be looked for in the Eternal Word alone. This St. Thomas lays down most distinctly. "Denominations are made from the more perfect. . . . But it is clear that the active and passive principles of generation of living things are seeds from which living things are generated. Therefore, Augustine calls conveniently all active and passive virtues that are principles of generations and movements, seminal reasons." [146]

The difficulty always returns to the same point. We do not say that the formative virtue in the seed is formally the seminal reason of the future being. Of itself it indicates only possible generation. The seminal reason is what is actually to exist in itself, now existing in its cause. We say that the seminal reason regards primarily the creative word, and is formally the abstract universal potency of prime matter determined by that word to this definite creature that according to the Creator's will, is to come into existence in its due form. In the order of nature, or of adminis-

[146] i, cxv, 2.0.

tration, the seminal reason is adequately the passive potency with the active potency that is to give it being; and because actual being implies the concurrence of so many causes and conditions, all of whatsoever kind that concur to any particular existence enter into its seminal reason. With actual existence the passive seminal reason becomes the existing thing, formally in its substantial form that actuates the determined potency of matter, the activity of which becomes a *partial* seminal reason of things future. Hence, though the formative virtue of the seed is of its *nature* higher and more perfect than the common virtues, as a *seminal reason,* considered apart from the passive seminal reason, it is not as complete as the common virtues active and passive working merely dispositively in their own order according to the determination received in the work of the six days. It contains the reason of the future being but partially, and cannot require the denial of the name to these virtues according to their share in the operation.

We may note once more the fundamental fact, that though St. Augustine uses the seed and its germination and fructification to illustrate his doctrine, he does not say that the seed is formally the seminal reason, or that the seminal reason is a seed. Behind the illustration may be seen always matter determined by the Creator's will to receive the seminal activity, without which actual germination or generation would be impossible.

Thus St. Thomas propounds his doctrine. It becomes his, because, having his own end in view, he proposes it in his own way; because, equal in intellect to the great Father of the Church, the greatest of Scholastics could not be a mere commentator; because of him not only can it be said, but must be said pre-eminently, that nothing passed under his hand without being left the brighter and clearer for his touch. But the substantial doctrine, whether read in St. Augustine or in St. Thomas, is absolutely the same, giving in neither any countenance to Evolution, whether Darwinian or Spencerian, or of any other form.

EPILOGUE

One thing now is certain, and it cuts the very
ground from beneath the feet of those asserting
the Evolutionism of St. Augustine. Seminal
reasons are not physical or mechanical or chemical forces. They are not energy, specific or particular, occasional or persistent, introduced into
matter to work out effects homogeneous or hetero-
geneous. St. Augustine never conceived them as
such; St. Thomas never understood them as such;
no one reading these Holy Doctors with the mini-
mum of decent respect can take them to be such.
They are of a higher order even than vital forces.
Natural History can be discussed adequately without an allusion to them. Creatures can be clas-
sified and named, divided into genera, and species, and families, without a suspicion of them
being excited in the naturalist's mind. But we
cannot cross the threshold of the Creator's tem-
ple to make the first inquiry into why these
alone exist of all possible creatures; how they
began, how they continue; without meeting the
seminal reasons face to face.

They are reasons, because they answer these
questions in the ultimate terms of the intelligence, the will, and the operation of the Creator, not in the secondary and instrumental terms
of physical agencies. They are seminal because

they contain what things are certainly to be, as the seed contains surely the future living being; because they contain them not actually, but virtually; and this by no native virtue, but by virtue of the creative word. On this point the whole doctrine turns.

St. Augustine would account for the origins of all material existence, not physically, or chemically, or biologically. Indeed, in this way the accounting would be but very imperfect; since, though physics, chemistry, biology precede certain individual origins, origins absolute of material things are necessarily antecedent to all such sciences. His discussion was rather metaphysical, not investigating the number, measure and weight of things, but rather the last reason why the actual number, measure and weight of things material should be just what it is, neither more nor less. Yet, not even the metaphysical question appealed to him directly. In the treatise *De Genesi ad Litteram* he sought the literal interpretation of God's infallible word; in *De Trinitate* he vindicated God's supreme dominion over every individual of the material creation, against any claim that might be made on behalf of spiritual creatures good or bad.

Fixed as the actual material creation is in number, measure and weight, to seek the last reason of this in matter itself would be folly. In itself matter is pure potency, negatively able to to be anything, with a possibility so universal, so equal, as of itself to be incapable of actuation. That horses and dogs exist actually and just so many of each kind, varying in size for each individual, calls for a positive determination of that negative passive potency, so that, no longer merely negatively determinable, it has a positive relation to just so many forms of individual horses and dogs that are to actuate it in future. This determination can come only from the creative act creating matter with such definite relations. That these relations should be realized fully by the actual existence of the successive individual horses and dogs, can come only from the same creative act, which, in determining matter, begins the existence of each; by bringing each into actual existence, accomplishes and perfects itself adequately.

Existence can be reached by the creature in two ways. It may be among the very first of its kind, the term of direct creation. If so, no other activity produces it than the creative act, which, first determining the potency of matter, then actuates that potency, and so has the existing creature as its adequate term. In the order of creation, therefore, nothing is to be found in matter as the seminal reason of the creature, but this positive determination of its purely passive potency to the absolutely certain future individual that will be in its existence the adequate term of the creative act. This seminal reason St. Augustine makes the formal term of creation in the day when the Lord made all things simultaneously.

Or it may come into existence by generation, or any other possible natural process. In this case some natural activity replaces the creative act as the immediate cause. But it cannot in the same way act as the sole efficient cause. To produce an effect by generation two things are requisite in matter that matter itself can not furnish. The first is the determination of the passive potency, we have just spoken of, to receive in this particular matter this particular form, by this particular generation, that is to say, the universal seminal reason necessitating the future existence of the creature, which is the foundation of its direct relation, as such, to the Creator. The second is the actuating of the active potency of the generator, enabling it by an added virtue to reproduce its own form in this definite matter. This again can come from no material force. We call it vital force, reproductive virtue, names that assert its existence, and its essential distinction from all other forces of nature. St. Augustine calls it the very hidden seed of seed, in which all fruitfulness is contained, a participation of creative power constituting the reproductive activity of the substantial form. Together with the determination of passive potency, it constitutes in the order of administration the usual seminal reason, both passive and active. Though, on account of the particular ends he had in view in the treatise, *De Genesi ad Litteram* and *De Trinitate,* St. Augustine confines his direct study of their seminal reasons to their relations to living beings, animal or vegetable, yet it is clear that his doctrine is to be extended, as St. Thomas extends it, to all that is physical in the order of

creation; to all that is physical and moral in the
order of administration.

In the actualization of seminal reasons according to the ordinary course of nature, though as
active potencies they are in material forces and
physical agents as in their subjects, these have dispositive functions only. This is seen daily in
generation, the germination of seeds, the operations of agriculture. What these agents do naturally, magicians may do with the aid of evil
spirits, and put the physical conditions of time,
place, conjunction, temperature, etc., for the production, say of serpents. The preternatural sudden existence and the equally sudden ceasing to
exist are effected by divine power, according to
the particular seminal reason decreed by the Creator's providence for this particular case. Of
miracles wrought by God through the ministry of
angels the explanation is essentially the same.
Should the miracle be wrought immediately, there
would be no question of any such disposition of
material elements and natural forces. The operation is analogous to creation. On the one side,
the obediential passive potency of matter determined in its first creation as the seminal reason
of this effect; on the other, the divine Word, supreme over every creature to which it responds
instantly.

This is St. Augustine's doctrine of seminal
reasons. Its end is not to give a theory of specific origins, but to explain as far as human mind
can penetrate it, the revealed truth of God's word,
the origin of all things by creation according as

the Sacred text narrates it, and God's continual operation in creatures, as the necessary, sufficient and sole reason of their existence and operation and propagation. In general, then, and so far as the order of administration is concerned, the seminal reasons may be considered as the term of the Creator's act, as such, constituting every being a creature in the strictest sense. They may be considered in the creature, where they are the immediate effect in material second causes of the unceasing operation of the First Cause moving all, giving all their formal efficiency. They may be considered in themselves. Then they are seen in every created being as in their subject. This subject, inasmuch as it exists and acts, may in its actual operation be called by a synecdoche, a seminal reason. But there is no identity. The seminal reason is the link binding the finite to the Infinite, the universe to God, so that He is ever the Creator, it, in its minutest element, its most insignificant phase always the creature. To make it a mere natural force or generating agent would lead to Pantheism rather than favor Evolution.

Imprimi potest.
F. C. Dillon, S. J., Præp. Prov. Californien.
Portlandii, Oreg., Aprilis, 14, 1923.
Nil obstat, J. M. Byrne, Censor Deputatus.
Imprimi licet, P. L. Ryan, Vicarius Generalis.
Sancti Francisci, Maii, 7, 1923.

www.ingramcontent.com/pod-product-compliance
Lightning Source LLC
Chambersburg PA
CBHW070446090426
42735CB00012B/2478